The War Cry in the Graeco-Roman World

This book aims to reconceptualise the Graeco-Roman military phenomenon of the "war cry"; the term itself is inadequate for defining an ancient military practice that has been misrepresented in modern media and understudied by contemporary scholars.

Gersbach introduces the term and paradigm "battle expression" to replace "war cry", which acknowledges the variety of undertakings, visual and sonic, that military forces from the Graeco-Roman world presented on the battlefield before, during or after battle. The "battle expression" was sophisticated in nature; it could include significant cultural song or dance that required high levels of rehearsal and execution. Conversely, battle expression types demonstrated spontaneous wit and humour on the part of a military force that aimed to capitalise on the experiences of a battle. These performances served a variety of purposes outside of instilling group cohesion among the participants and to intimidate the onlooking enemy. This book associates the psychological dimension of warfare, religious identity and military strategy supported by the High Command to this practice. In addition, the author draws comparisons with later historical periods, as well as the actions of modern-day European football supporters in stadiums, to reconstruct the atmosphere created by ancient military forces on the battlefield.

The War Cry in the Graeco-Roman World is suitable for students and scholars of Classical Studies, particularly those interested in ancient warfare and military history, as well as those studying the history of warfare more broadly.

James Gersbach is a PhD graduate from Macquarie University Sydney, Australia. He is the head teacher of the history department at Secondary College in Sydney, Australia. This is his first major publication.

Routledge Monographs in Classical Studies

For more information on this series, visit: www.routledge.com/Routledge-Monographs-in-Classical-Studies/book-series/RMCS

The War Cry in the Graeco-Roman World

James Gersbach

Routledge
Taylor & Francis Group

LONDON AND NEW YORK

First published 2023
by Routledge
4 Park Square, Milton Park, Abingdon, Oxon OX14 4RN

and by Routledge
605 Third Avenue, New York, NY 10158

Routledge is an imprint of the Taylor & Francis Group, an informa business

British Library Cataloguing-in-Publication Data
A catalogue record for this book is available from the British Library

Library of Congress Cataloging-in-Publication Data
Names: Gersbach, James, author.
Title: The war cry in the Graeco-Roman world / James Gersbach.
Description: Abingdon, Oxon ; New York, NY : Routledge, [2023] |
    Series: Routledge monographs in classical studies | Includes bibliographical
    references and index.
Identifiers: LCCN 2022030750 (print) | LCCN 2022030751 (ebook) |
    ISBN 9781032248585 (hardback) | ISBN 9781032248608 (paperback) |
    ISBN 9781003280439 (ebook)
Subjects: LCSH: Military art and science—History—To 500. | Battle-cries. |
    Greece—History, Military—To 146 B.C. | Rome—History, Military.
Classification: LCC U33 .G694 2023 (print) | LCC U33 (ebook) |
    DDC 355.00938—dc23/eng/20220921
LC record available at https://lccn.loc.gov/2022030750
LC ebook record available at https://lccn.loc.gov/2022030751

ISBN: 978-1-032-24858-5 (hbk)
ISBN: 978-1-032-24860-8 (pbk)
ISBN: 978-1-003-28043-9 (ebk)

DOI: 10.4324/9781003280439

Typeset in Times New Roman
by Apex CoVantage, LLC

Contents

Acknowledgements

This book would not have been possible without the support and encouragement from family and friends. The mentorship and insightfulness of Associate Professor Peter Keegan, Macquarie University, Sydney, has been instrumental in the formation of this work. The love and understanding given by my wife, children and parents ensured that this study was undertaken and completed. A driving force behind it all was the life of "Salvo" Susino.

Thank you.

1 From war cry to the "battle expression"

What is a war cry?

The war cry from the ancient past is understood by contemporary society as being simple and primitive in noise and purpose. Modern media forms, such as film and television programmes, have depicted war cries from the ancient world, as performed *en masse* by military forces that stare at each other across the battlefield for the purpose of heightening tension on the cusp of battle and displaying the combatants' preparedness. War cries are presented as unsophisticated in nature; they are predominately inaudible yells and screams that may contain the clashing noises of weapons and shields. The war cry is portrayed as mass noise with little sophistication, variation or meaning. Moreover, there is an absence of any substantial scholarly depth study that focuses holistically on this military phenomenon. There has been no acknowledgement that the study of war cries reveals unique cultural identifiers as well as an exhibition of homogeneous practice. This has negatively impacted on our understanding in the modern era of the nature and purpose of war cries from the Graeco-Roman world. The modern understanding of the term war cry does not reflect the ancient practice that is presented in ancient Graeco-Roman literary works and archaeological remains. Significant modern reference material that relates to the Graeco-Roman world has little to no definition of the term.[1] This reflects the broader lack of any substantial study, within modern scholarly works, regarding it.[2] However, even a cursory review of the extant sources referring to the diverse contexts wherein this ancient military phenomenon can be situated indicates the need for a new explanatory paradigm.

According to Graeco-Roman literary sources, the nature and purpose of war cries was diverse. Ancient sources ascribe specific characteristics to battlefield performances undertaken by different cultural groups throughout the Mediterranean world. Admittedly, these characteristics, or stereotypes, provided by Graeco-Roman authors seem to be based upon the racial attitude of the Greeks or Romans towards themselves and the cultural group referred to in the literary source. Non-Graeco-Roman military forces that are referred to in literary sources serve either as an ally unit within a Greek or Roman military force, or they are the direct opponents. Often these cultural characteristics are politically driven as a means of bias

DOI: 10.4324/9781003280439-1

and propaganda to demonstrate the alien customs of foreign people in comparison to the superior or civilised Graeco-Roman culture.[3]

An entirely different understanding of the ancient war cry needs to be formed, which acts in contrast to the modern-day perception of it. Ancient historical literary works, whose authors were eyewitnesses to or had access to eyewitness accounts concerning military engagements and forces, in conjunction with other ancient literary forms from the Graeco-Roman world, such as ancient play scripts, poems and hymns, suggest that an ancient war cry, in the modern sense, should be viewed in a different light. The literary evidence reveals that the war cry was a means for ancient military forces to express themselves in a variety of ways – to communicate and to express identity, religious and political belief, humour, musical achievement and unit cohesiveness to the enemy and themselves – and the purpose of the expression was significant to different cultural backgrounds. The manner that military forces from the Graeco-Roman world communicated aspects of their culture to the enemy, and re-affirmed it to themselves, was just as varied as the expression itself. A combination, or selection, of vocal noise, bodily movement, music and silence could suffice. The modern understanding of an ancient war cry does not reflect what the expression of a military force in antiquity was nor does it accurately explain how it was communicated. Therefore, to move away from the stereotype of a war cry, understood from the modern perspective as simplistic and unsophisticated noise and movement of a military force before, during and/or after battle, a new definition needs to be applied. This ancient military phenomenon will be disassociated from the general term "war cry" and, as applied to pertinent episodes in the extant sources about the ancient Graeco-Roman world, identified instead within the broader conceptual frame of the *battle expression*.

The definition of a battle expression encompasses any performance, individual or *en masse*, by any member/s of an ancient military force from the Graeco-Roman world. This performance consisted of visual and/or sonic displays that aimed to have an overwhelmingly positive impact on those who undertook it and a negative impact on the enemy. The battle expression could comprise intentional silence, noise (shouting, singing, clapping, clashing arms together, replicating animal noises, playing musical instruments, pre-battle speeches), visual presentations (helmet plumes, uniform, shield decoration, war paint, hairstyles, standards) and/or physical action (dancing, rhythmic jumping, waving, swaying, shaking). Whatever the form the battle expression took, Graeco-Roman authors suggest they could be performed spontaneously or rehearsed by a military force. Graeco-Roman authors advocate they could take place before, during and/or after a military engagement, such as a battle or siege, by a military force, often united in noise, visuals and/or movement.

The term "battle expression" evokes a more genuine reflection of what the source material promotes about group cohesion and intimidatory practices of ancient Mediterranean military forces. Moreover, the frame of the battle expression model attempts to recognise the unique and various methods employed by military forces to communicate to the enemy and to themselves – as revealed through the perspectives of Graeco-Roman authors. The significance of the methods used by different military forces is, similarly, not limited to the idea of battle expression. The term war cry, and the modern-day interpretation of it, does not acknowledge that the methods used could be significant to the religious, cultural or political background of a military force. Battle expression allows for a fresh and holistic interpretation and understanding of an ancient military phenomenon that does not have the feature, of massed unsophisticated noise suggested by modern usages of war cry. In fact, many types of battle expression, as will be further discussed, contained little vocal noise, but consisted more of musical and/or united movement performed by military forces. Further, the manifestation of the war cry, from a modern-day perspective, is associated with feelings of hostility and aggression, whereas literary evidence suggests ancient military forces expressed sentiments of worship, joy, enthusiasm and other emotions besides anger, hatred, and hostility before, during and/or after the battle. The term battle expression more suitably reflects an ancient military phenomenon that has otherwise been misinterpreted.

The translation of ancient literary works from the original Latin and Greek supports the need for a greater understanding of the Graeco-Roman war cry. For example, Henry George Liddell and Robert Scott's *A Greek-English Lexicon*[4] provides evidence that the term war cry is deficient in its role as a descriptor of ancient military practice. Throughout Liddell and Scott's lexicon, there are 12 separate entries that refer to the English term war cry. Within these entries, the term war cry is used as a translation for a variety of different actions and emotions that ancient Greek authors used to refer to the methods by which ancient military forces communicated with the enemy and/or themselves. The variety of different terms used in this lexicon to refer to a war cry reveals that the English definition for the term does not relate to the ancient meaning. The war cry is defined in lexicographical terms as a phrase or name shouted to rally troops and the modern representation of this is through unsophisticated noise. However, in this lexicon, a war cry is linked to Greek terms that are also defined in the following ways: to shout the shout of victory, to call on the god of the war cry, loud united noise which is comparable to animal noises, such as a flight of birds, a signal or watchword to begin a battle and emotions linked to dread, pain and joy.[5] Clearly, the English term for this ancient military phenomenon is too general and simplistic to account for the diversity of this ancient military phenomenon. The term battle expression would better suit as a translation for the various Greek terms used in Liddell and Scott's lexicon that is referred to as a war cry. The definition of battle expression acknowledges that ancient military forces adopted a diverse array of methods for expressing themselves, such as shouts or statements, animal noises and cries based on different emotions. Also, through the term battle expression, the significance of what ancient military forces expressed is acknowledged. Religious sentiment is a theme in many battle expression forms from the ancient Mediterranean world and battle expression attempts to recognise that ancient military forces communicated significant cultural features, such as religious belief, on the battlefield. For example, in Liddell and Scott's lexicon, a Greek term that evokes the benevolence of a god is listed under the translation of war cry.

The adoption of the battle expression paradigm has been influenced by the content of ancient literary works and modern equivalents that reflect similar characteristics. Four notable factors underpin the formulation and development of the battle expression. Each factor provided an understanding of the purpose, nature and impact of ancient battlefield practices that battle expression encapsulates. Firstly, the numerous types of battle expression references found within the works of Graeco-Roman authors, such as Arrian, Caesar, Xenophon and Ammianus Marcellinus, highlight culturally homogenous and heterogenous traditions. These practices include praising and worshipping individual leaders/heroes/deities, expressing the founding sociopolitical origins of a military force and taunting the opposition with insults and jeers. These were expressed by military forces through creating noise and/or movement *en masse*.

Secondly, the battle expression phenomenon has manifested itself in medieval, pre-modern- and modern-day contexts. Medieval transcripts and historical works reveal that French military forces would sing the *"Song of Roland"* before the battle. This has been associated with the Norman army prior to the commencement of the Battle of Hastings in AD 1066.[6] Archaeological evidence from the AD 15th century, for example, demonstrates that familial war cries were prevalent centuries after antiquity – that can be applied to the battle expression definition.[7] So, too, battle expressions were performed by combatants in modern theatres of military conflict,[8] namely the American Civil War (Rebel Yell), the Gallipoli campaign during the First World War ("Heads up to the Warwicks!") and World War Two Nazi followers exclaiming "Sieg Heil". Modern-day U.S. marines universally exclaim "Hoo-rah" as their cry of association. New Zealand sporting teams, particularly its national rugby teams such as the All Blacks, perform the "Haka" before a match. The Haka is a traditional Polynesian battle expression that is not solely reserved for the Maori of New Zealand. Haka attempts to unite the group undertaking it – by inspiring them to achieve great feats and reminding those performing it about their identity, people and land. In a competitive/military environment, Haka serves to intimidate those to whom it is directed. Protest movements have demonstrated elements of the battle expression phenomenon through collective chants of protesters, depending on the movement itself, and may demand the resignation of political leaders, the protection of environments or a moratorium on military action.

Thirdly, although manifest in a competitive, rather than military, context, the variety and purposes of football chants performed by club supporters,[9] especially those aligned to (but not limited to) teams competing in Germany and England, within purpose-built stadiums correlate favourably to the performances of the battle expression described in the ancient sources. The relationship between battle expression from the ancient Graeco-Roman world and the songs and chants of European football supporters in modern sports stadiums is evident in the German term for football chants or showing support for one's football team. This term is *Schlachtgesänge*, which literally translates into English as "battle songs". In this sense, the tradition of singing, chanting and moving in unison, performed by a body of people or individually, within an agonistic, intensely competitive environment, has continued into modern society from the ancient past to the present day. The types of themes used in football chants in the modern day can be used as a comparison to the types of themes used for the battle expression from the ancient Graeco-Roman world. Individual football players, both past and present, are honoured just as past and present warriors, such as generals, kings and heroes, both divine and mortal, were honoured in antiquity. Football supporters commonly sing and chant about their socio-political origins; for example, the English football team "Queens Park Rangers" has the supporters chant "We are QPR, we are QPR", and we can also note the iconic Liverpool FC anthem of "You'll Never Walk Alone". Graeco-Roman armies also communicated to their opponents, through chanting and singing, where they were from and who they were, for example, the Ambrones tribe.[10]

Football supporters of the modern era often taunt their opponents through chants and songs that have been used in the past and rehearsed, but taunts can be created and performed spontaneously depending on the circumstances of the football match. Similarly, ancient armies used taunts to unnerve the enemy and distract them from their orders.[11] Ancient references to military performances in a battlefield environment – armies clashing their weapons together and stamping their feet on the ground, as well as jumping rhythmically into the air – resonate with the fanatical performances rehearsed and enacted across European, North, Central and South American, Asian and Australasian football stadiums of the modern era. Instead of brandishing weapons at opposing groups of supporters (even though weapons are still carried by many football hooligan supporters throughout Italy and Argentina), however, hand clapping is used. The popular and iconic "Viking Clap" performed by Icelandic football supporters in the UEFA Euro 2016 tournament is a notable example. This supporter action has since been adopted by the supporters of other "Viking"-affiliated sporting teams around the world since, such as the Canberra Raiders in the NRL and the Minnesota Vikings in the NFL.

Lastly, ancient military battles have been a topic of great interest among historians and the public over many centuries. Media forms, such as literature, artwork and film, have depicted ancient military battles as both gruesome and heroic and have used battles from antiquity to entertain, educate and attract audiences.[12] Similarly, ancient historical accounts, poems and plays have based their works on battles and wars from their recent past and antiquity for identical purposes.[13] Such works include Homer and his account of the Trojan War, Aeschylus and the Persian Wars, Caesar and his commentaries on the Gallic and Civil Wars, and Tacitus' biographical *Agricola*, to name a few. The medieval age was an era that passed on works from antiquity through monks copying and preserving ancient texts in monasteries. Medieval poetry and plays have done likewise, recording and incorporating ancient battles into entertainment, notably the legend of King Arthur and Shakespeare's historical plays, such as *Antony and Cleopatra*. Modern-day media such as Hollywood blockbuster movies have used the contexts of ancient wars and battles as climatic scenes within the subtexts of the plots to films. Popular movies such as *Spartacus, Ben-Hur, Alexander, Gladiator, Centurion, King Arthur, The Eagle, Troy* and *300* have all attempted to incorporate ancient military battles and wars into their productions to attract the masses. Within human nature, there appears to be a strange attraction to warfare, particularly from the ancient world. The intrigue of the lost societies of Rome and Greece and their martial nature continue to attract widespread interest amongst the peoples of the modern world.

Because of this popularity, the modern world has created many popular historical works and much research has been undertaken to account for, explain, describe, analyse and evaluate the intricate facets of ancient warfare, particularly with respect to the societies of classical Greece and Rome. Numerous volumes of scholarly journal articles, histories and reference materials have been created that reflect the fascination people have with this ancient human phenomenon.[14] Aspects of ancient warfare such as tactics and strategies employed, uniforms and armour worn, weapons used, training undertaken, disciplinary codes, recruitment methods, hierarchy and officer classes of the various peoples of the ancient world, to list a mere few, have been published on a large scale. Television documentaries have been created using archaeological inquiry to determine and unravel the true course, and nature, of ancient military battles.[15]

Despite this interest, very little research exists regarding the ancient battle expression as a distinct phenomenon and practice. According to the ancient literary record, that which originates, particularly, from the Graeco-Roman world, the battle expression appears to have been a fundamental part of military life. Julius Caesar in his *Civil Wars* refers to the long-established practice of military forces in antiquity adopting, encouraging and implementing battle expression before, during and after battles:

> Between the two lines just enough space had been left for a charge by each army. But Pompey had instructed his men to absorb Caesar's charge; they were not to move from their position, but to allow his line to break itself up. People say that he did this on the advice of Gaius Triarius, so that the soldiers' first powerful charge would be rendered ineffective and Caesar's line distended, and that his own men in proper formation could attack a scattered enemy. Pompey also hoped that the spears would fall more lightly on soldiers held in place than if they themselves ran into projectiles coming at them, likewise that Caesar's soldiers would be disheartened by the double run because they would also be undone by exhaustion. But to me at least this action seems to have been taken by Pompey for no valid reason, because there is a certain stirring of the spirit and an eagerness naturally inborn in all men that is kindled by enthusiasm for combat. Commanders should not repress this but augment it. Nor is it a pointless ancient institution that battle signals sound from all sides and every man raises a shout. By these things, they thought, the enemies are terrified and one's own men incited.[16]

Caesar identifies here the fundamental nature of the battle expression within the context of the ancient Mediterranean battlefield. As a side note to his historical account of the battle of Pharsalus (48 BC, against the forces of Pompey the Great), Caesar explains that Pompey's tactical order prior to engagement with the enemy, namely, for his troops to be silent and still, was flawed. Caesar claims that military generals should encourage their troops by any means to extract spirit and keenness for battle. He observes that raising a battle expression puts fear into the enemy's hearts but also instils motivation within the men performing the cry. An interesting point raised by Caesar in this same extract is his acknowledgement that the practice of performing battle expression was ancient in his day and had been in use ever since for sound reasons.[17] Caesar's statements – made most probably to discredit Pompey for failing to follow simple protocol in battle, or perhaps to reflect poorly, in comparison to his own superior tactics, on Pompey's approach to this important military engagement – suggest that it was uncommon for armies of the ancient Mediterranean *not* to perform battle expression before, during or after military engagements. Therefore, we may understand Caesar's implication that battle expressions were, in fact, typical features of military life as significant support for this study's contention that they represented a continuing facet of military practice (*antiquitus institutum est*). As a result, in-depth research into this topic would shed valuable light on an issue otherwise only considered *en passant* in the scholarship of military history in antiquity.

The importance of the battle expression in the military life of the peoples from the ancient Graeco-Roman world is evident in many ancient works. The question, then, arises as to why so little research into understanding it has been undertaken by modern scholars.[18] If there is strong and consistent evidence of the battle expression being a fundamental aspect of ancient military life in antiquity and scholars, students and enthusiasts from the modern world are so attracted to the various military practices, aspects and features of ancient military forces, why has this topic been so consistently overlooked and ignored?

Ultimately, the answer to this question resides in the problematic nature of the ancient evidence. The difficulties associated with examining the phenomenon of the battle expression are evident in the ancient authors' lack of detailed recording. The reason for this lack of detail can be divided into several categories. Firstly, the purpose of the ancient authors was not to describe in detail the precise sequence of actions forming the performance of a battle expression. Perhaps the audience the author was writing to did not need to be educated on the phenomenon due to their familiarity with or existing knowledge and experience of the practice, as could very well be the case for the literate members of Graeco-Roman society who traditionally held positions of leadership and importance within their respective military forces.[19] Alternatively, perhaps most authors did not want to educate their audience in any detail about a phenomenon that, in relation to the typical genre where such encounters were recorded (namely, historical narrative), was nothing more than a colourful way to bring a dramatic battle scenario to climax. Importantly, many ancient authors may not have first-hand experience of a battle expression and therefore may not have any points of reference for understanding the physical, acoustic or verbal "vocabulary" of it. On the other hand, those authors who were present at the military engagements their historical works describe (e.g. Caesar, Xenophon, Ammianus Marcellinus, Josephus) will not always have been able to witness the performance in its totality due to obstructions at the battle. So, too, those authors who were able to provide eyewitness testimony may not have remembered the actual wording, action or sound of the phenomenon because of poor or incomplete memory or length of time passed between the military engagement and the writing process. Naturally, those authors who were not actually present at the military engagements described in their narratives (e.g. Arrian, Tacitus) will not have been able to provide the details of the phenomenon.[20]

An example where the ancient author has obstructed a clear understanding of a battle expression due to a lack of specific detail is evident in Sallust's *Jugurthine War*. During Sallust's preamble to a military engagement, a Numidian force, consisting of Numidians, Mauretanians and Gaetulans allied to Jugurtha, had surrounded a Roman force on top of two hills. As day turned to night, the African force, having surrounded the Romans, began shouting and singing through the night in anticipation of the next day's battle.[21] Sallust claims this practice was customary of these "barbarians",[22] emphasising the otherness or cultural *difference* of foreign battle expression compared to Roman. From an African perspective, the tradition of making loud noise during the night before battle must have been significant and appears to have been a normal feature prior to battle. The songs, hymns, shouts or dances that would have been performed are not mentioned; however, the subject of the songs must have had a culturally familiar or resonant religious, mythological, historical or social undertone for the majority, if not all, in the composite African force to participate. According to Sallust, the battle expression of specifically Numidian and broader African military forces, contrary to the Roman perspective, were both geographically and ethnically contextualised *and* sophisticated, suggesting a strong cultural connection or familiarity amongst most of the participants to the battle expression performed.

This evidence is simply too vague to allow us to say much of anything beyond speculative argumentation. Sallust's inability to provide specific details regarding this battle expression resides in Sallust not being a contemporary of the conflict he recorded.[23] Despite Sallust's knowledge of military affairs and his intimate relationship with Africa, the purpose of composing this history was not to highlight the military practices of Romans or Numidians. Sallust aimed to present to the audience both the capabilities of Marius and the incompetence of the traditional Roman elite.

While these factors invariably complicate the validity, reliability and broader utility of the surviving source material, the lack of detail in the literary record that treats the phenomenon can be mitigated effectively through other patterns of research. The use of archaeological evidence – for instance, *glandes* or sling bullets used in ancient military engagements – may help alleviate the problematic nature of the contemporary historical sources associated with this phenomenon. The discovery of sling bullets and other missile objects used by military forces during battles and sieges appears to have been, on certain occasions, inscribed with messages and/or symbols prior to firing. In most cases, *glandes* display inscribed names of military leaders (generals and kings) with honorific titles, such as Cnaeus Magnus Imperator[24] and King Philip II of Macedon.[25] Sling bullets also commonly portrayed symbols associated with the act of striking, such as a spear head, scorpion or eagle.[26] Other sling bullets contained inscribed exclamations that mocked or taunted the enemy with suggestive statements of inflicting pain.[27] Finally, other *glandes* bear the name of the military force from which the slinger originated, such as the legion number or unit name. These inscriptions on sling bullets could possibly contain specific wording, mottos, noises or actions of battle expression from specific military forces. As a result, inscribed *glandes* comprised a very particular element of battlefield engagement that may be seen to accompany – or, in various contexts, may well have constituted a specific component of – the battle expression of certain military forces.

Iconographic evidence discovered on ancient monuments, such as Trajan's Column and the Arch of Titus, reveals important features of ancient military life that support literary evidence or provide added insight not contained within the literary source material. Numismatic evidence, such as coins minted during Constantine I's reign that depict Roman military standards, sheds further light on significant features of the ancient phenomenon beyond the literary sphere.

Literary sources other than strictly historical accounts can shed light on the exact wording, sound and action of a battle expression. Poetry and plays from the Graeco-Roman world contain many references to the phenomenon, including descriptions regarding the performance and wording. Poets and playwrights such as Homer, Aeschylus and Tyrtaeus specifically refer to the wording of battle expression and the actions performed by the participants. For example, Aeschylus' *Agamemnon* describes the sound of the Argives' battle expression as the sound of eagles screaming.[28] Similarly, in *Persians*, the Greeks are said to have raised a battle expression that struck terror into the enemy forces.[29] So, too, the war songs of Tyrtaeus outline in detail the wording of military elegies which were performed by Spartan hoplites.[30]

While close critical examination of a spectrum of extant literary, epigraphic and material data reveals that armies, particularly those of Greece and Rome, did in fact employ battle expression on a large scale, the phenomenon has a limited presence in current scholarship.[31] This is surprising given the number of references in ancient sources. The battle expression and the constituent elements of collective verbal and physical expression (dance, clapping, stamping, raising arms, clashing weapons) performed by a military force, such as taunts and victory songs, all appear to serve the same purpose. The purpose was to solidify a military force together, to instil fear/intimidation into the enemy, to build up the confidence of the military force performing it, to invoke the spirits of ancestors/deities and to demonstrate origin, or past deeds. Ancient writers who used sources of information from accounts contemporary to, or eyewitness testimony from those present at, the given battles, do provide greater detail into the typology, range and function.[32]

The ancient Graeco-Roman battle expression appears to have been taught, rehearsed, trained and practiced on a large scale.[33] Songs of worship and taunts were fundamental to military life. Naturally, any indication of such a typical feature of ancient military practice provides greater insight into social customs, traditions and declarations of intent or purpose associated with the phenomenon. Further, establishing a typology of battle expression reinforces ideas already known from the ancient world regarding religious observance, cultural and racial beliefs and individual/collective psychology in times of military conflict. Alternatively, a detailed study of the typology and range unearth otherwise unknown information regarding the importance of music, religion, dance, poetry and/or local custom in military contexts.

The following chapters in this book create a refreshed outlook of the ancient battle expression. Each formulated chapter from this study has been inspired by the source material which suggests that the modern-day understanding of the term is a misrepresentation of an ancient military phenomenon. According to literary and archaeological evidence, the battle expression phenomenon contained universal features that appear consistent across cultural groups within the Graeco-Roman world. Ancient military forces rehearsed and prepared their soldiers for the effective execution of battle expression types and to psychologically prepare their men for the enemy's attempts to intimidate them through the same processes. The aims and purposes of implementing a battle expression on the battlefield were universal, irrespective of culture. Whether this tradition was encouraged from the top down, through sponsorship from the high command, or if it was controlled at the "grassroots" level of the army, geared from the rank and file to influence the high command is debatable. The evidence suggests a predominantly top-down influence.

Chapter 2 will explore the chronological sequence of the Graeco-Roman authors and pertinent archaeological material that relates to the battle expression. Authors who were contemporary to wars and battles that they record generally provide more detail in this field and are, hence, more useful than authors who wrote centuries after the events they described. This does not mean that these latter authors are not useful in studying the battle expression, as their own understanding of the battle expression from their time period is imposed upon the events they ascribe which reveals much about the phenomena. The exploration of the authors' perspectives will highlight that African, Asian, Celt/Germanic, Greek and Roman battle expression types differed and were presented by the authors as unique in their own way. The authors' presentations were based on patriotism, xenophobia, political propaganda and depth of understanding.

Chapter 3 focuses on the impressive atmospheric conditions that battle expressions created on the battlefield. Experimental archaeology can be applied to reconstruct the visual and sonic nature of battle expression types in the Graeco-Roman world. Useful comparisons found in the actions of European football supporter groups, specifically, inside stadiums aid in reconstructing the atmospheric conditions that would have been reminiscent on ancient battlefields. The sound produced by hundreds and thousands of troops on a battlefield through song, clashing weapons, playing instruments, jumping in unison, clapping or mimicking animals was impressive. The familiarity military forces had with what they undertook contributed to its effectiveness in sound and appearance. Military forces exploited the landscape to maximise the sound that could be generated with their battle expression, valleys contributed to heighten the noise produced. Alternatively, some Graeco-Roman military forces intentionally used silence to generate an unnerving atmosphere that displayed tremendous levels of discipline and rehearsal. The battle expression aimed to produce an atmosphere that was friendly to the participant and distracting to an enemy.

Chapters 4–8 highlight the commonalities that existed between the battle expression types of ancient military forces of the ancient Graeco-Roman world. Chapter 4 reveals the role the battle expression played in strengthening the psychological resolve of a military force prior to battle. Military forces used massed movement and/or noise to develop and gauge group cohesion and battle readiness of the men. These practices aimed to boost confidence levels amongst the fighting group and would consist of choreographed or spontaneous performances. Types of battle expression that served to achieve these aims revolved around the familiar cultural song and dance (if the military force shared a similar culture), humour and wit – often at the expense of the enemy. Chapter 5 focuses on the use of the battle expression to psychologically impair the enemy through intimidatory practices that served to erode the enthusiasm levels of the enemy for battle, mainly through taunting. The potential for the battle expression to distract, impose fear and unnerve an enemy prior to battle was a key component to their use and had links to overall military strategy.

Chapter 6 details the religious dimension of the battle expression. Military forces and individuals within them often turned to patron deities and religious tradition to find resolve and courage in the lead up to the battle. The glorification and invocation of divine forces were common features of religious-inspired battle expression types. Examples from the ancient literary and archaeological source material are used to highlight these purposes. Chapter 7 demonstrates that another aspect of battle expression was extolling socio-political and military identity. Military forces took pride in professing the origins of their socio-political identity to the enemy by way of uniformed appearance, praising political leaders and recounting social customs. Similarly, military leaders and units were praised for their fighting qualities and past deeds to prompt the fighting group to emulate them. Examples from the ancient literary and archaeological source material are used to highlight these purposes.

Chapter 8 explores the custom of oathtaking in a military context on the battlefield, with a focus on Greek and Roman tradition. Ancient literary sources record the practice of commanders of Graeco-Roman military forces swearing a public oath or making a covenant with specific deities prior to the battle for the purpose of gaining the support of the chosen deities in the upcoming battle. Entire military forces could take part in this oath making their actions in the battle accountable to the deities. Often the oaths or covenants required the commemoration of the deity in some physical form, such as the construction of a temple, or the regular worship of that deity. Examples from the ancient literary and archaeological source material are used to highlight these purposes.

Notes

1 Such as the *OCD* ed. Hornblower, Spawforth & Eidiner (2012); Sabin & de Souza (2007): 399–460 (*Hellenistic world and the Roman Republic); Gilliver* (2007): 122–157 (*Late Republic and the Principate)*; Rance (2007): 342–378 (*later Roman Empire*).

2 Authors such as Krentz (1985); Speidel (2004); Rance (2015); Cowan (2007); Hanson (1989); Pritchett (1971) and Baray (2014) have produced work that refers to the "war cry" of specific ancient cultures such as Germanic, Celtic, Roman and Greek, but there is a lack of a holistic study of the ancient "war cry".

3 Modern scholars acknowledge the tendency of Graeco-Roman authors to characterise foreigners/barbarians as alien for the purpose of sponsoring their own civilization. This may explain the cultural stereotypes associated with battle expression in the ancient literary sources.

   For example, see Grant (1995): 67–74; Marincola (2009): 17–18; Baynham (2009): 290; Feldherr (2009): 302–303.

4 Liddell & Scott. (1940).

5 1.ἀλα^λ-άζω . . . *raise the war-cry . . . shout the shout of* victory, 2.ἀλα^λαί . . . exclam. of joy . . . *god of the war-cry,* 3.ἀλαλή . . . *loud cry . . .* esp. *war-cry . . . battle,* 4.ἀλα^λητός . . . *shout of victory . . . war-cry, battle-shout,* 5.ἀναβο-άω . . . *cry, shout aloud,* esp. in sign of grief or astonishment . . . the *war-cry,* 6.ἀν-α^λα^λάζω . . . *raise a war-cry,* 7.ἐλελεῦ . . . *a cry* of pain . . . a *war-cry,* 8.ἐνοπή . . . *crying, shouting,* as of birds . . . esp. *war-cry, battle-shout,* 9.ἐπα^λα^λάζω . . . *raise the war-cry,* 10.σημεῖον . . . *signal for battle . . . watchword, war-cry,* 11.στονόεις . . . *causing groans* or *sighs* . . . (war-cry), 12.συνεπ-α^λα^λάζω . . . *join in raising the war-cry.*

6 William of Malmesbury. *Gesta regnum Anglorum* Bk 3.

7 Grancsay (1931): 14. *"Io Harr" ("I persevere")* ten times as the family war cry in battle.

8 For example, "Attack and Die" by McWhiney & Jamieson (1984).

9 In this instance, football refers to soccer, however, the atmosphere generated by fans consisting of purposeful action (vocally, intentional movement and appearance) can be associated with a host of sports such as NFL, NBA, AFL and NRL to name just a few.

10 Plut. *Sul.* 18.3.

11 Caes. *B Civ* 1.69, 3.48. [Caes.] *Spanish Wars* (30–31). NB The commentary on the *Spanish Wars* is not thought to have been written by Caesar, but rather (perhaps) by Aulus Hirtius.

12 For example, we find the popularity of modern films which contain recreations of ancient battles; *"Gladiator"* (2000) $187, 705, 427 Box Office takings within the U.S, *"300"* (2007) $210, 614, 939 Box Office takings within the U.S.

13 For more on this topic, see the *chapter "When War Is Performed, What Do Soldiers and Veterans Want to Hear and See and Why?"* From Palaima (2014).

14 Such as: Greece & Rome; Historia; *The Cambridge History of Greek and Roman Warfare*, Vols. I and II; Sabin & de Souza (2007); Hanson (1989); Macdonald (1992); Pritchett (1971); Pushkin & Elton (1934); Sabin (2000); Worthington (2008); Baray (2014).

15 Such as the television series: *Battlefield Detectives*, The History Channel (2003–2006) & *Battlefield Britain*, BBC (2004).

16 Caes. *B Civ.* 3.92.

17 Caes. *B Civ.* 3.92.5.

18 Sabin (2007): 401. The study of the war cry phenomenon should be considered as a key feature of the "face of battle". For further reading see regarding the "face of battle" see Keegan (1977).

19 For a brief overview of Roman historians and their audience, see Marincola (2009): 12–15.

20 Sabin (2007): 399.

21 Sal. *Jug.* 98.6–7.

22 Sal. *Jug.* 98.6.
23 Sallust wrote the history of the war against Jugurtha ca. 40BC whilst the war ended towards the end of the 2nd century BC, ca. 106 BC.
24 Ariño (2005): 234.
25 Foss (1975): 28.
26 Foss (1975): 27.
27 Foss (1975): 28.
28 Aesch. *Ag.* 49ff.
29 Aesch. *Pers.* 384ff.
30 See Bayliss (2017) & Banks (1853): 327–343.
31 Authors such as Speidel (2004); Rance (2015); Cowan (2007); Hanson (1989); Krentz (1985) and Pritchett (1971) have produced work that refers to the war cry or its associates such as shouting before battle of specific ancient cultures such as Germanic, Roman and Greek, but there is a lack of any holistic study of the ancient war cry.
32 Namely, Ammianus Marcellinus, Arrian, Caesar, Livy, Josephus, Plutarch, Tacitus and Xenophon.
33 Amm. 22.4.6; Jos. *BJ*: 3.70ff.

2 Graeco-Roman literary and archaeological sources

A chronological survey of the Graeco-Roman authors used in this book will provide greater understanding and context for the information about the battle expression that they record. In the first instance, authors will be categorised into groupings based on the time during which they wrote. From this, a general overview of the literary works that refer to battle expression types will be highlighted to determine the relationship the author had with the subject they described; what is known about the intimacy the author had with the events; and if any literary formula or style was in trend during that period. Secondly, assessing where appropriate the influence on literary texts of the socio-historical and cultural contexts within which the corpus of source material relevant to the battle expression was produced will be an essential component in what follows. Pertinent archaeological source material will be integrated into the relevant time periods in conjunction with the literary authors. Definitions for "the Romans" and other peoples referred to in these sources will serve to clarify and unpack the composition of military forces mentioned. In combination, these points of reference will establish the effective limits of the evidentiary foundation on which the arguments presented in this book are based.

Before moving forward, it is important to note that the intention of this survey of the source material is to profile the backgrounds, descriptive strategies and literary techniques employed by relevant authors only insofar as they frame the validity, reliability and usefulness of surviving primary and secondary written texts which speak to the phenomenon of the battle expression. In doing so, surveying the time when authors produced their works will highlight the prevalence of the battle expression paradigm across the Graeco-Roman world. Irrespective of whether authors had a political, moral or literary agenda when creating their works, what will become clear is that the frequency and diversity of literary references to the battle expression – often found in relation to the creation of a climactic battle narrative – reveals that authors and their audiences were familiar with them. Archaeological remains have been incorporated into this study to supplement the literary source material for evidence relating to battle expression. The study of glandes, or sling bullets, provides valuable insight into the nature of battle expression types as well as understanding the psychological intent of specific military units in battle. The reading of inscriptions and analysing the images engraved on

DOI: 10.4324/9781003280439-2

sling bullets, that have been uncovered at sites of ancient sieges, have revealed important information pertaining to battlefield preparation and custom for associated military units. Iconographical evidence, such as the Alexander mosaic from Pompeii and Trajan's Victory Column in Rome, provides images that signify the appearance of ancient armies and military units adopted on the battlefield as a form of battle expression. The collaboration of the literary and archaeological evidence elucidates that the phenomenon was significant and commonplace in the military life of the Mediterranean world throughout the period in which the select authors lived and wrote.

Authors that pre-date the 5th century BC, including Homer and Tyrtaeus, were generally poets who devised epic and foundational stories for their societies. There is great conjecture amongst modern scholars over their reliability as they were often not contemporaries to the events they recorded; they tend to sensationalise events and descriptions; they make generic suggestions. For example, Homer in his *Iliad* refers to Trojan military practices, such as battle expression, however, he was not a contemporary. As a result, Homer ascribes the Trojans with known Greek battle expression types such as assigning Greek deities and their characteristics to Trojan ones which may reveal more about the contemporary Greek battle expression than *other* cultural groups. Homer's portrayal of warfare during the Iliad's Trojan war reflects the contemporary style of warfare practiced during Homer's lifetime rather than from the period in which he refers.[1] Despite these shortcomings, early authors collectively record battlefield practices that fall within the battle expression paradigm. Pre-5th-century BC works reveal that battle expression existed as a military practice that was prevalent in those societies, particularly in the Greek world from where these authors derived. Battle expression types recorded in these literary works are supported by later Graeco-Roman authors who reaffirm the themes and lyrics of Tyrtaeus' war songs in Sparta.[2]

Graeco-Roman authors from the 5th to 4th centuries BC consisted of Greek historians and playwrights. These authors can be considered quite reliable when studying Greek military practices from their written works as they were experienced in military matters. Thucydides and Xenophon, for example, possessed intimate knowledge of the events they wrote about and military affairs in general. The writing techniques of Thucydides and Herodotus have come under much scrutiny by modern scholars due to Thucydides' invention of speeches and Herodotus' moral lessons that drove his work. Nevertheless, Herodotus interviewed eyewitnesses to military events he described, and Thucydides had a military career and would have been privy to the military customs of his day.

When referring to the battle expression, playwrights such as Aeschylus and Aristophanes provide no elaboration surrounding the practice to the audience. This suggests that both author and audience (the citizen body of Athens) would have been familiar with the battlefield customs presented in the plays. It would be an expectation that a playwright details at length pre- and post-battle undertakings of an army, such as the singing of the paean, if the author or audience were not familiar with it. This is not the case in Greek plays from this time period, suggesting both playwright and audience understood battlefield processes and customs.[3] The references would not be integrated within the plays if they were not completely relevant or accurate. On this matter, Aeschylus was a participant in the Persian Wars, namely at Marathon and Salamis, and would have been familiar with Athenian military customs. Aristophanes, on the other hand, did not have a military career but lived and wrote plays during the Peloponnesian War. The frequency of military engagements throughout this period meant that he was acquainted with military matters through eyewitness testimony from the citizen population. Aristophanes' audience would have contained citizens who had participated in the war and expected references to military custom in plays to be relevant and familiar.

Fourth-century BC sling bullets uncovered from Philip II of Macedon's siege of Olynthus provide insight into the intentions and nature of Macedonian battle expression during this siege. The inscriptions discovered on lead sling bullets fired from the Macedonian forces contain statements that evoke their intentions at the siege. Sling bullets used at Olynthus record recurring messages of "conquer" and "blood".[4] Macedonian battle expression served to instil the units within the army with a sense of determination for victory and the drive to inflict devastation onto the enemy as evident with these sling bullet remains. Besides this, Macedonian sling bullets showcased a sense of humour through the messages they contained, alluding that their battle expression forms were aimed to amuse the units within it through the nature of the custom. Messages of "ouch", "take it" and "a candy"[5] appear, which present a level of entertainment and enjoyment amongst the participants when undertaking these practices.

Polybius is categorised as a 3rd–2nd-century BC author who was self-aware of what history should achieve and the qualities a historian needed to have. He believed that history was didactic and opposed sensationalism. He professed to be concerned only with recording the truth in political and military matters.

> A historical author should not try to thrill his readers by such exaggerated pictures, nor should he, like a tragic poet, try to imagine the probable utterances of his characters or reckon up all the consequences probably incidental to the occurrences with which he deals, but simply record what really happened and what really was said, however commonplace. For the object of tragedy is not the same as that of history but quite the opposite. The tragic poet should thrill and charm his audience for the moment by the verisimilitude of the words he puts into his characters' mouths, but it is the task of the historian to instruct and convince for all time serious students by the truth of the facts and the speeches he narrates, since in the one case it is the probable that takes precedence, even if it be untrue, the purpose being to create illusion in spectators, in the other it is the truth, the purpose being to confer benefit on learners.[6]

Polybius was intimate with military affairs, he was well-travelled and was present at key military engagements he recorded, such as the final siege of Carthage ca.149 BC.

The Alexander Mosaic from the House of the Faun in Pompeii originates from around the 2nd–1st centuries BC. This mosaic depicts a scene that contains Alexander the Great with some of his Macedonian armies in battle against the Persian King Darius III and a group of his men. The dramatic scene presented in the mosaic showcases the moment in battle when Darius began to flee from the field in a chariot out of Alexander's reach. The mosaic is styled off a 4th–3rd-century BC Hellenistic painting and the image was converted into mosaic form after Roman forces brought the original painting back to Rome as booty after 2nd-century BC military campaigns in the Hellenistic East. The mosaic reveals the divergent appearance of Macedonian and Persian military forces on the battlefield. Studying the visual appearance of ancient armies has revealed insight into the nature and purpose of battle expression forms, namely the importance of identification on the battlefield and association with specific fighting units. In the Alexander mosaic, the companion cavalry is depicted with their unique helmet type.

The 1st century BC–AD 1st century has the largest number of Graeco-Roman authors that refer to battle expression types. Some authors from this period wrote about events well before their contemporary society, such as Livy and Plutarch, whilst others recorded events that transpired during the time and were eyewitness testimonies, such as Caesar and Josephus. The author Varro wrote about the Latin language and within this work instances of battle expression were discovered that highlighted the impact this military phenomenon had within ancient society aside from military contexts.[7] The amount of military-oriented texts produced during this time period reflects the militaristic nature of society and the changing political environment of this time – which was accustomed to war and military matters. This was a period of great transition especially in the Roman world as it expanded its influence around the Mediterranean and developed under the rule of Augustus and his successors. Many of these written works record the military exploits of Rome and the achievements of influential and powerful men of these formative centuries.

Roman glandes dated to the Perusine War from 41 to 40 BC provide evidence for the continuity of the military tradition of inscribing images and messages onto sling bullets for use in battle. The recurring messages from the inscriptions demonstrate the intensity and aggression that battle expression forms could adopt in the lead up to and during battle. Evidence of taunting and crude sexual metaphors appears on sling bullets. Often to the detriment of individual figures of an enemy force, military units hoped to inflict pain and suffering on the enemy while being entertained in the process.[8] The discovery of sling bullets from this period reveals their continued military use from the 4th century BC with the continued practice of inscribing messages, and images on them elucidate their importance in understanding the mood and intent of the soldiers who fired them at the time.

Authors from the AD 2nd to 3rd centuries had military careers or were privy to source material that originated from military campaigns and eyewitness testimony. Tacitus, Arrian and Dio Cassius were authors from this time period that lived military careers and had access to eyewitness testimony for the topics they wrote about. Other authors from this period, such as Polyaenus, Suetonius and

Athenaeus, were not as well acquainted with military life as the aforementioned authors, however, relied on source material that was reliable. The iconographic reliefs visible on Trajan's triumphal column from Rome originate from the AD 2nd century and record Trajan's Dacian military campaign. Scenes presented on this column depict the military appearance, such as uniforms, weapons and armour used by the various military units within the Roman and Dacian armies. The intentional appearance created by military units served, among other things, for identification purposes and was linked to battle expression custom. Like the study of the Alexander mosaic, this artwork was created by contemporary artists that represented different military units based on their appearance.[9] Trajan's monumental column served the purpose of elevating the emperor to the heavens to praise his deeds, in doing so, the depiction of events and those involved, such as the military forces, required accurate representation.

Ammianus Marcellinus and Vegetius are notable authors from the AD 4th to 5th centuries. Both wrote about military matters, however, Ammianus was the most experienced of these authors due to his role in the Roman army during the reign of Roman Emperor Julian. Ammianus' participation in Roman campaigns in Germany and Syria is particularly useful in the investigation of the battle expression as he recalled these consistently in his battle narratives. Vegetius' work on Roman military affairs was inspired by the racial and military contamination of the Roman army during this period where he saw the need to return to the military ways of previous centuries. Despite Vegetius' lack of military career, he became an expert in Roman military life from the research he undertook to prepare for his book *Rei Militaris Instituta*.

The Roman Emperor Maurice or another general of his reign has been credited for the writing of the *Strategikon* in the 6th century. Whoever the original author of this book was, it was composed by an experienced military careerist, which gives weight to its reliability and usefulness for its references to the battle expression. Procopius is another notable author from the 6th century who had a significant military career. He accompanied Belisarius on military campaign during this period. Notable historians from this time had great familiarity with military matters. Their literary works demonstrate the continuing presence the battle expression had at this late stage of antiquity.

Graeco-Roman authors impressed cultural characterisation onto the battle expression types they recorded. The cultural profiling of ancient armies and their battle expression is a common feature found in the source material that relates to the battle expression. Each military culture in the Graeco-Roman Mediterranean world adopted unique forms to transmit their battlefield customs. The cultural background of the military force influenced the nature of battle expression types that exist in literary evidence. Ancient authors focused on the cultural differences between armies to convey the author's political agenda. Roman historians, particularly Livy and Caesar, characterised *other* cultural groups in comparison to the Romans for socio-political agenda.[10] Ancient authors presented African, Asian, Celt/Germanic, Greek and Roman battle expression types as different and unique in their own way. The authors' presentations were based on patriotism, xenophobia, political propaganda and depth of understanding.[11] The following text contains an overview of the cultural typecasting that authors linked battle expression with. It is important to be aware of these prejudices and characterisations when analysing the evidence when deciphering the intentions, nature and impacts of the battle expression irrespective of which culture undertook it.

The battle expression of African military forces, namely Numidian and Carthaginian, is represented by ancient Graeco-Roman authors as quite peculiar in nature and purpose. The evidence reflects the conception of the African continent and its people as a multi-racial entity suggesting a diversity in noise, language and customs amongst the various African peoples. The unique examples of African battle expression, compared to other ancient Mediterranean cultures recorded in Graeco-Roman historical works, are derived from the languages of North Africa. The Afroasiatic languages that make up the lands of Africa that border the Mediterranean are just as diverse and inter-mingled with different dialects and influences in the modern world as they would have been in the ancient world.[12] The linguistic and habitual diversity of African military forces would have served to generate this characterisation of African battle expression found in the historical sources.

The Roman historian Sallust in his work *The Jugurthine War* is one of the more insightful ancient literary sources that detail the battle expression of the Numidian culture. Numidia, during the republic, was an ally kingdom of Rome in North Africa. However, as a result of the hostile actions of one of its leaders, Jugurtha, Rome and Numidian forces, led by Jugurtha, became embroiled in a long-drawn-out war, which resulted in Rome's ultimate victory during the late 2nd century BC. The kingdom of Numidia, after this war, was reinstated as an ally kingdom of Rome.

Graeco-Roman literary sources present African military forces' battle expression as peculiar in nature. Their inclusion in the historical record of the Romans in particular (Livy and Sallust) highlights the notion that the African military culture was different and used by authors as interesting anecdotes for their audiences. The African battle expression culture is characterised by attempts to deceive or deter the enemy from their military objectives, such as the Ligurians in Livy.[13] Also, African military units aimed to inspire their compatriots through familiar movement and sound that was alien to the author,[14] particularly evident in Sallust, resulting in the various episodes being included in historical works intended for Graeco-Roman audiences. The characteristics of African battle expression reinforce the notion that their military culture was presented as unique in the Mediterranean world.

Ancient Graeco-Roman literary records characterise Asian battle expression as ostentatious displays that reflected the cultural diversity of the peoples within this geographical region. Asian military forces are presented in the literary record as effectually arrogant, demonstratively flamboyant and psychologically egotistic in their battle expression. This makes the Asian battle expression unique[15] when compared to other cultures around the Mediterranean world. Trojans, Lydians, Scythians, Persians and Parthians are all categorised for this study as Asian cultures that fought directly against, or within, the confines of Graeco-Roman military units from the ancient Mediterranean world. Evidence for Asian battle expression types is consistently found within Graeco-Roman literary accounts dated across the broad time span of this study. The presence of Asian battle expression is found in Homer, Herodotus, Xenophon, Arrian, Plutarch and Ammianus Marcellinus.[16] From the Trojan War (12th–8th centuries BC) through to the late Roman empire (AD 4th–5th centuries), a variety of different Asian military forces are presented as undertaking a battle expression that represents the traditional characteristics associated with the Asian military culture. This suggests, primarily, that Asian cultures were stylistically stereotyped by Graeco-Roman authors during this period, but also, were an ever-present military participant in the historical interplay of Greece and Rome in the Mediterranean and Eurasian antiquity.

The sole evidence available for the Asia Minor culture of archaic Troy derives from the Greek Poet Homer and his epic poem the *Iliad*. This poem's depiction of Trojan battle expression contains culturally conflicting traits. On the one hand, Homer ascribes distinct individualism to certain forms of Trojan battle expression, separating them from the range of expression characterising the Greeks' tactical approach to military engagement. Conversely, Homer represents certain forms of Trojan battle expression as very similar in nature to the Greek battle expression performed during the same war. Here, Homer denotes the Trojan battle expression as the army sounding and moving like a flock of birds.[17] Homer compares the sight and sound of the Trojan army gathering and presenting themselves on the battlefield as watching a flock of birds – *like* geese, cranes or swans, gathering on an Asian meadow in immense numbers. The sound that the army generated made the earth echo.[18] Homer likens the appearance of the mentioned birds as being very bold and the sound of the flock/army crying together as overwhelming.

The battlefield customs of Asian military forces are presented in Graeco-Roman literary sources with unique cultural attributes that separated them from others in the ancient Mediterranean world. No doubt Graeco-Roman authors intentionally shaped this image for an audience accustomed to conflict with the powers of the oriental east. Within the manipulated image, the Asian battle expression can be viewed as empowering and meaningful to the participant. The sophistication of sound and action within Asian battlefield custom is acknowledged through the concept of battle expression. Excessive noise and numerical superiority associated with Asian armies inspired flamboyant and arrogant displays. The public acclamation of one man being the king of all kings and the exhibition of slain enemies' heads provides insight into a culture that was presented as highly confident, yet callous. Sonic and visual displays that reflected movement through colour and ostentatious sound aimed to overwhelm the enemy before the battle began. This is suggestive of a culture intimate with deception. The incorporation of drums and stringed instruments, within a military context, highlights the wealth and extravagance of the Asian culture.

Graeco-Roman literary sources portray both ancient Celtic and Germanic battle expressions as reflective[19] of the natural surroundings that this culture was accustomed to. For the purpose of this study, the military forces of the Celts and

Germans will be merged together when focusing on their battle expression. The differences in geographical and political origins among the various Celtic and Germanic tribes referred to in Graeco-Roman literary works are not intended to be overlooked or ignored.[20] It is through the nature, range and purpose of Celtic and Germanic battle expressions that their cultures are linked. Ancient Celts and Germans were animists.[21] The raw power of nature found in water, animals and weather was revered by the ancient Celts/Germans. The physical world of the Celts and Germans was inextricably connected to their spiritual and cultural world that was full of supernatural forces.[22] It was their animism that was integrated within their battle expression.[23] It is this cultural feature that differentiates Celtic and Germanic tribes from other ancient Mediterranean cultures. The literary sources present Celtic and Germanic military forces as intimidating in appearance and atmospheric in noise.[24] Through appearance and noise, the Celts and Germans sought to extract the raw power of natural forces and assault the senses of the enemy with them. The battle expression of the ancient Celts and Germans testify to the inextricable relationship this culture had with nature and the forces they observed within it that their warriors attempted to mimic on the battlefield.

Ancient Celtic and Germanic battle expression types are recorded in detail by Graeco-Roman authors. This is not surprising given the long history of Greek and Roman military forces that have been in direct military contact with Celtic and Germanic tribes. The Romans had military contact with Celtic and Germanic tribes since the 4th century BC and continued to fight with, and against, these forces up to the fall of Constantinople in AD 1453. The northern borderlands of both Greek kingdoms/city-states and Roman provinces were occupied and raided by Celtic and Germanic tribes throughout their histories. The northern and western expansion, and later the defence, of the Roman empire from the 2nd century BC through to AD 5th century was focused upon the lands that the Celts and Germans occupied. The detail offered by ancient Graeco-Roman authors, such as Livy, Ammianus Marcellinus, Plutarch and Tacitus,[25] regarding the battle expression of the Celts and Germans, was an established writing style and a staple feature of battle narrative when accounting for the military exploits of the barbaric northern tribes.

The appearance of Celt/German warriors before and during battle is described by many Graeco-Roman authors at great length in their histories. The dedication authors give to this subject suggests that their audience took great pleasure in reading these accounts. Of all the cultures that had military contact with the Greeks and the Romans, the descriptions of the Celt/German warriors by Graeco-Roman authors excel all others in detail and imagery. This reveals how alien Celt and German culture was to the Graeco-Roman, reflected notably in their battle expression. The appearance of Celtic and Germanic military forces instilled fear and overawed their opponents. The wild, natural bearing of these warriors sought to intimidate the on-looker. Polybius presents a large Celtic army, during Rome's Cisalpine Gaul campaign against the Celts in the 3rd century BC, unnerving the Roman army sent to destroy it. Polybius claims that the sight of the Celtic battle expression, which involved simultaneous movements of the warriors who were naked and in excellent physical condition struck fear into the Romans.[26] Many Graeco-Roman authors testify to the bare appearance of Celtic and Germanic warriors. The contrast between the Greek/Roman warrior, heavily clad in armour (helmet, breastplate, shield), to the Celt/German, who was naked, except for weapon, shield and potentially a piece of jewellery such as a torque, must have been memorable to the eyewitness. This is potentially the reason for Celt/ German warriors being referred to by many different Graeco-Roman authors.[27] The demonstration of tall/large physiques and naked bodies not only would have been intimidating to Greek/Roman warriors who were not generally as large compared to the Celt/German peoples,[28] but this practice was potentially a reflection of the Celtic/Germanic animistic culture that embraced natural power. A Celtic/ Germanic military force baring their naked bodies, and emphasising their naturally large physique, to the foreign Roman/Greek enemies may have served to unite the warriors through cultural identity.[29]

The purposeful creation of a physical appearance that reflected a natural force/s to intimidate the enemy is a typical cultural military trait unique amongst the ancient Celts/Germans. The ancient literary sources categorise the battle expression of the Celts and Germans with descriptions of nudity, the manipulation of hair, the dying of the body and shields to reflect night/hell, the flashing of teeth, the mimicking of wild animals, leaping into the air *en masse* and displaying of large weapons. These actions each demonstrate the natural force Celts and Germans tried to emulate on the battlefield, strength. These warriors intentionally presented themselves as stronger, or more powerful than their opposition through their natural bearing and their attempts to emulate forces from nature that were more powerful than them.

The evidence from ancient Graeco-Roman literary sources, together with Celtic/Germanic artworks and early middle-age literature of Celtic/Germanic warriors in battle, dispels the sentiment of ancient battle expression being primitive and barbarous. Through interpreting the sources, Celtic/Germanic battle expression types are portrayed as being rooted in meaning, purpose and honour. Animistic displays, such as altering behaviour and appearance to mimic animals, natural forces and deities, were deeply religious and culturally important. Ritualistic dancing, the profession of patriotism and character traits of the clan/people evoked sophistication and ancestral pride. The nature of Celtic/Germanic battle expression highlights the close affiliation the military had with the culture of these peoples. A new interpretation of the sources suggests that, for the Celts/Germans, war facilitated the worship of their supreme deities and of nature.

Graeco-Roman literary sources present Greek battle expression as reflective of the socio-political customs and religious beliefs of the poleis. In contrast to other ancient Mediterranean cultures, the types of battle expressions employed by Greek armies were controlled and cohesive undertakings.[30] Greek military forces undertook calculated battle expression forms that sought to honour the deities and proudly proclaim socio-political customs. Evidence suggests that all Greek military forces, irrespective of which polis or geographical region the force originated from, consistently adopted, and performed the paean[31] or religious hymn before the battle. Literary evidence reveals that the battle expression of armies from different regions in the Hellenistic world, such as Sparta and Macedon, reflected their socio-political identity. Greek military forces displayed these characteristics from the pre-archaic through to the Hellenistic era of the 2nd century BC, an extensive military tradition.

Greek armies customarily undertook battle expression. The paean hymn was universally sung by Greek armies before battle (often in conjunction with blood sacrifice) and after victorious battle.[32] The pre-battle paean was the traditional precursor to the *eleleu* cry that Greek armies utilised in the opening stages of battle that evoked martial deities, suggesting a religious dimension to the tradition. Archaeological and literary evidence reveals that Greek city states purposefully affirmed their socio-political ideologies on the battlefield to instil resolve amongst the army, whilst the nature of the battle expression aimed to create an atmosphere nerve-wracking for the enemy. Spartan armies integrated the war poetry of Tyrtaeus and customs founded by Lycurgus into their battle expression.[33] Theban armies revealed their connection to Herakles through painted motifs on their shields.[34] Macedonian armies professed their admiration and loyalty to Philip II through their incorporation of the sarissa into their battlefield customs.[35] The inscription of Philip's name on lead sling bullets, discovered at Olynthus, reveals the proud connection that the Macedonian military had to their monarchical leaders in a military context.[36]

The literary tradition records the phenomenon of the battle expression (from the republic to the late empire). As we have seen in relation to other ancient military cultures, alterations in form of battle expression developed over time, in this instance, as Rome's dominance extended across the Mediterranean world. While early and mid-republican Rome encompasses a period of significant political and military change, it is also important to note that the term "Roman" should be understood to encapsulate the peoples of Italy integrated within the Roman military. Evidence suggests that typical forms of Roman battle expression, practiced during the republic and early empire, were still in use during the late empire.[37] The literary record portrays a shift in forms of battle expression during Rome's imperial period reflecting the integration into the Roman army of non-Italian soldiers (as compared to previous periods), and, in consequence, elements of non-Roman battle expression. This shift resulted in the Roman army adopting Germanic and Eurasian forms of battle expression as well as a range of Christian military invocations at the expense of traditional Roman forms that embodied pre-Christian ideology.[38]

The forms of battle expression employed by the Roman army were traditional and long-established practices. The clashing of weapons against shield and massed vocal noise, which potentially incorporated song or chant, originated from archaic times.[39] Significant religious customs influenced battle expression; the *Salii* customs and military triumphs provide insight into the origins and cultural connection religion had with military custom. Roman battle expression embraced unique stimuli that served to inspire friendly troops yet terrifies the enemy. The blasting of trumpets, aside from their practical function of issuing orders, was integrated within the battle expression. Military standards, such as the eagle, were looked upon by Roman troops and the enemy with awe.[40] Roman sling bullet inscriptions support the unique features of the Roman battle expression.[41] Civil war periods created awkwardness on the battlefield when Roman legions fought against each other. Roman civil war periods highlighted the significant cultural dimension that battle expression embraced. The development and diversification in types of Roman battle expression resulted from the civil war periods. The influence for this change came from the integration of non-Roman units in the army.[42] The adoption of non-Roman military practices, such as the *barritus* and *draco*, highlights the cultural shift in the nature and customs of the army. Ultimately, Roman religious customs and beliefs were central to their battle expression. It is not surprising that the rising influence of Christianity during the late empire saw Roman armies replace traditional pagan types in favour of Christian equivalents.[43]

The study of the authors and archaeological evidence that relates to the battle expression is necessary to determine the reliability of the source material. It is evident that the authors who were eyewitnesses or contemporary to military events that they present in their works provide an accurate image concerning battlefield customs. It is apparent that the representation of the battle expression, in terms of the nature and purpose, over the course of centuries remains consistent within the source material. Authors who had eyewitness testimony available to them as source material are equally well versed in understanding their subject. What is clear is that the battle expression has been recorded by Graeco-Roman authors from before the 5th century BC through AD 6th century. The authors range from historians, playwrights, chroniclers and artists. They all testify to this paradigm. Often the source material highlighted the cultural diversity between ancient armies for political agenda and used the battle expression to reinforce this. Irrespective of the cultural profiling that took place in Graeco-Roman literary and iconographical works, the battle expression is presented as multi-faceted in nature, typology and purpose. The cultural characterisation found within the source material reveals an added element to our understanding of the practices of this military phenomenon.

Notes

1 For more on this, see Van Wees (1994) "*The Homeric Way of War: The 'Iliad' and the Hoplite Phalanx (II)*" Greece & Rome, Vol. 41, No. 2 (Oct., 1994), pp. 131–155. & Vol. 41, No. 1 (Apr., 1994), pp. 1–18.
2 Ath. *Dei.* 14.630–631.
3 For more on this topic, see the *chapter "When War Is Performed, What Do Soldiers and Veterans Want to Hear and See and Why?"* From Palaima (2014).
4 McDermott (1942): 36–37; Foss (1975): 28.
5 McDermott (1942): 36–37; Foss (1975): 28.
6 Polyb. 2.56.10–13.
7 Varro. *DL.* 5.73; 6.68; 7.49.

8 Kelly (2012): 291–294; McDermott (1942): 36–37.
9 Speidel's work "*Ancient Germanic Warriors: Warrior Styles From Trajan's Column to Icelandic Sagas*" London: Routledge, 2004, is based on this notion.
10 Vasaly. (2009): 245–260.
11 For example, see Grant (1995): 67–74; Marincola (2009): 17–18; Baynham (2009): 290; Feldherr (2009): 302–303; Lendon (1999): 275, 280–281; Glück (1964): 25.
12 Blench, R. (2006) 'Chapter 4: Afroasiatic' pp. 139–162.
13 Livy. 35.11.6–11.
14 Sal. *Jug.* 60.3–4.
15 Hom. *Il.* 2.459–469, 3.1–5, 5.595–597; Hdt. 1.17, 3.151; Xen. *Ana.* 1.7.4, 1.8.11; Xen. *Cy.* 3.2.9–10, 3.3.58–63; Arr. *Ana.* 4.4.2, 4.18.6; Plut. *Sul.* 16; Plut. *Cras.* 26; Amm. 19.1.8, 19.2.6, 19.2.11–12, 20.7.5.
16 Hom. *Il.* 2.459–469, 3.1–5, 5.595–597; Hdt. 1.17, 3.151; Xen. *Ana.* 1.7.4, 1.8.11; Xen. *Cy.* 3.2.9–10, 3.3.58–63; Arr. *Ana.* 4.4.2, 4.18.6; Plut. *Sul.* 16; Plut. *Cras.* 26; Amm. 19.1.8, 19.2.6, 19.2.11–12, 20.7.5.
17 Hom. *Il.* 2.459–469.
18 Hom. *Il.* 2.466.
19 By "reflective" means the ancient authors' descriptions of Celtic and Germanic battle expression demonstrate their animism. The ancient authors did not intend to understand or interpret the cultural connection of the Celts and Germans to nature themselves; their accounts reinforce this understanding.
20 Sometimes it is difficult to determine the differences between Celts and Germans from the ancient Graeco-Roman world. See Baray (2014) and Rawlinson, C. "*On the Ethnography of the Cimbri*" The Journal of the Anthropological Institute of Great Britain and Ireland, Vol. 6 (1877), pp. 150–158. See Strab. *Geo.* 4.4.2.
21 Speidel (2002): 276; Luc. *Phar.* 1.450–455; Tac. *Ger.* 9–10; Strab. *Geo.* 4.4.4.
22 Tac. *Ger.* 9–10, 45; Luc. *Phar.* 1.498–501.
23 Speidel (2004): 1, 13, 43, 47, 51, 57 & 81.
24 Polyb. 2.28; Livy 38.17ff; Dio. Cass. 38.45.4–5.
25 Polyb. 2.28; Livy. 38.17; Tac. *Ger.* 43; Amm. 16.12.43.
26 Polyb. 2.29.
27 References for Celt/German warriors' fearsome physique and/or naked appearance in battle include; Tac. *Hist.* 2.22; Polyb. 2.28–29; Livy. 38.17; Caes. *Gal.* 2.30.
   For a study on the biological standard of living in Europe, see Koepke & Baten (2005).
28 Caes. *Gal.* 2.30.
29 Speidel (2002): 276.
30 Hom. *Il.* 3.1–9; Ath. *Dei.* 14.624; Xen. *Hel.* 4.3.17; Pritchett (1971): 108; Potter (1728): 84.
31 *OCD* 4th ed. (2012): 1060 "*Paean*"; Strabo. *Geo.* 9.3.10–12.
32 Pritchett (1971): 105; Rutherford (1994): 113–116; Haldane (1965): 33 n.5; Potter (1728): 76; Thuc. 1.50, 2.91, 4.43, 4.96, 7.44; Arr. *Ana.* 1.15.7–8; Xen. *Ana.* 1.8.16–19, 4.3.18–19, 4.3.29, 4.3.31, 4.8.16, 5.2.13–14; Xen. *Hel.* 2.4.17, 4.2.19; Xen. *Cy.* 3.3.58; Aesch. *Pers.* 384–395.
33 Bayliss (2017); Cartledge (2006): 79; Kõiv (2005): 238, 263; Ath. *Dei.* 14.630ff; Plut. Cleo. 2.3; Plut. *Mor.* 959a.
34 Xen. *Hell.* 7.5.20.
35 Arr. *Ana.* 1.6.1–4.
36 McDermott (1942): 36–37; Foss (1975): 28.
37 Cowan (2007); Rance (2015).
38 Cowan (2007); Rance (2015).
39 Caes. *B Civ.* 3.92; Cowan (2007); Dusanic (2003): 91.
40 Livy 28.14.10; Amm. 27.2.6, 28.5.3, 29.5.15; Jos. *BJ.* 3.123; Tac. *Ann.* 2.17; Speidel (1984): 17–22.

41 Kelly (2012): 291–294; McDermott (1942): 36–37.
42 Amm. 16.12.43; Dio. Cass. 72.16.1–2; Tac. *Ger.* 3; Tac. *Hist* 2.21, 5.16–17; Speidel (1984): 118–128. For research surrounding recruitment of the Roman army, see Dobson & Mann (1973); Littleton & Thomas (1978): 520 & Wadge (1987): 209. For a study on the Germanic warriors in the Roman army depicted on Trajan's column, see Speidel (2004).
43 Cowan (2007); Rance (2015).

3   Battlefield atmosphere

Battlefield terrain could influence the selection, and impact, of a battle expression. Ancient military forces exploited the landscape of a battlefield to maximise the sound that could be generated with their battle expression. For example, valleys and hills contributed to prolonging and heightening the noise produced. Alternatively, some Graeco-Roman military forces intentionally used silence to generate an unnerving atmosphere that displayed tremendous levels of discipline and rehearsal. The battle expression aimed to produce an atmosphere that was friendly to the participant and distracting to an enemy. The sound produced by hundreds to thousands of troops on a battlefield through song, clashing weapons, playing instruments, jumping in unison, clapping or mimicking animals was impressive. The familiarity military forces had with what they undertook contributed to its effectiveness in sound and appearance. Experimental archaeology can be applied to reconstruct the visual and sonic nature of battle expression types in the Graeco-Roman world. Useful comparisons found in the actions of European football supporter groups, specifically, inside stadiums aid in reconstructing the atmospheric conditions that would have been reminiscent on ancient battlefields.

Ancient literary sources reveal that the terrain of a battlefield could be intentionally used by the military high command to heighten the volume of noise emitted by a battle expression to manufacture an electric atmosphere prior to battle. This atmosphere of intense and overwhelming sound served to impose dread and anxiety onto the enemy forces to gain a military advantage that could influence the outcome of the engagement. Geographical features, such as hills and valleys, created optimal conditions for increased levels of noise generated through vibrating sound off landforms making an echo effect. Frontinus, in his Strategems, refers to a method that involved using the surrounding terrain that consisted of hills to deceive the enemy into believing that there was a larger force gathered. The use of the hills to exacerbate the sound of blaring trumpets that echoed the noise instilled fear in the enemy that resulted in them fleeing in terror.

> The general Minucius Rufus, hard pressed by the Scordiscans and Dacians, for whom he was no match in numbers, sent his brother and a small squadron of cavalry on ahead, along with a detachment of trumpeters, directing him, as soon as he should see the battle begin, to show himself suddenly from the

DOI: 10.4324/9781003280439-3

opposite quarter and to order the trumpeters to blow their horns. Then, when the hill-tops re-echoed with the sound, the impression of a huge multitude was borne in upon the enemy, who fled in terror.[1]

Frontinus is clear in his assertion that general Rufus ordered this battle expression. The military high command aimed to utilise the hills that comprised the battlefield to create such an atmosphere of noise to represent a force larger in size than what was visible. Interestingly, the result of this planned battle expression was that it avoided bloodshed and resulted in a strategic victory for Rufus' force. The notion that battles in the Graeco-Roman world could be decided on the basis of the undertaking of a battle expression is presented in this source, further supporting the idea that the war-cry term is insufficient in its definition to represent the performances that took place on battlefields in ancient times.

Arrian presents an episode in his Anabasis where Alexander the Great instructed his phalanx to perform a combination of spontaneous and rehearsed battle expression in mountainous terrain. The aim of this was to force the enemy out of an advantageous position of higher ground. Arrian is adamant that the high command ordered this undertaking that served to integrate training drills and traditional battle expression. Arrian claims Alexander ordered his phalanx to maintain strict silence and then, when ordered, to perform numerous training drills that required the phalanx to move their sarissa spears in unison, by lowering, raising and swinging them in various directions. The intention of this was to demonstrate the skill and discipline of the Macedonian phalanx in front of the enemy. In the setting of the mountainous terrain, the sound generated by massed sarissa spears moving in unison would have made an impressive noise, echoed by the hills. After this demonstration of skill, Alexander ordered the traditional vocal battle expression with clashing of spears on shields to be undertaken. Arrian claims that it was at this point that the enemy force fled from their advantageous strategic position that commanded the high ground.

> In the circumstances Alexander drew up his phalanx with a depth of 120 files. On either wing he posted 200 horsemen, bidding them keep silent and smartly obey the word of command; the hoplites were ordered first to raise their spears upright, and then, on the word, to lower them for a charge, swinging their serried points first to the right, then to the left; he moved the phalanx itself smartly forward, and then wheeled it alternately to right and left. Thus, he deployed and maneuvered it in many difficult formations in a brief time, and then making a kind of wedge from his phalanx on the left, he led it to the attack. The enemy, long bewildered both at the smartness and the discipline of the drill, did not await the approach of Alexander's troops, but abandoned the first hills. Alexander ordered the Macedonians to raise their battle-cry and clang their spears upon their shields, and the Taulantians, even more terrified at the noise, hastily withdrew back to the city.[2]

Literary evidence suggests that battle expression types were intentionally utilised by military forces to exploit the battlefield terrain to achieve desired outcomes.

These outcomes included the profession of military identity onto an enemy to achieve atmospheric supremacy over them, or as an extension of military strategy. Ammianus records the battle expression of an AD 4th-century Roman army and their Persian opponent that was extended by the surrounding hills of the battlefield terrain.

> the hills re-echoed with the shouts which rose on either hand. Our men extolled the prowess of Constantius Caesar, 'lord of all things and of the world', while the Persians hailed Sapor as Saanshah and Peroz, titles which signify 'king of kings' and 'conqueror of war' . . . at a call from the trumpets, the battle was renewed.[3]

Ammianus explicitly records a competitive monomachy of noise between the Roman and Persian forces during this engagement. Both sides used a traditional battle expression that glorified their socio-political identity as a tool to gain atmospheric dominance over their rival before battle was renewed. What is interesting about this recorded battle expression is the reference to the surrounding terrain being a factor in the creation of this loud battlefield atmosphere. Ammianus suggests that both Roman and Persian forces realised the impact the hills had on the noise produced when the battle expression was performed. Both forces aimed to exploit this feature to achieve sonic dominance on the battlefield prior to physical conflict.

Through Livy's description of the Battle of Lake Trasimene, fought between Carthaginian and Roman-led forces in 217 BC, the landscape of the battlefield seems to have been intentionally used by Hannibal to communicate to the Roman army that they had been outmanoeuvred. According to Livy, Hannibal exploited the surrounding hills of the battlefield in unison with the low-lying mist that covered the lake and plains where the Romans were gathered to launch a charge with the performance of a battle expression.

> The Phoenician had now gained his object, the Romans were hemmed in between the mountains and the lake and their escape cut off by his own troops, when he made the signal for all his forces to attack at once. As they charged down, each at the nearest point, their onset was all the more sudden and unforeseen inasmuch as the mist from the lake lay less thickly on the heights than on the plain, and the attacking columns had been clearly visible to one another from the various hills and had therefore delivered their charge at more nearly the same instant. From the shouting that arose on every side the Romans learned, before they could clearly see, that they were surrounded; and they were already engaged on their front and flank before they could properly form up or get out their arms and draw their swords.[4]

Livy suggests that Hannibal devised his army's charge to surprise the Romans. In doing so, Hannibal maximised the battlefield terrain to screen his troops from the Romans but stationed them so that his troops were visible to each other – positioned on the hills above the mist. When Hannibal signalled for the attack to begin, his army is recorded to have struck up a vocal battle expression that Livy claims informed the Romans that they had been surrounded. The noise generated by the hills and across the lake reached the Romans from different directions as the charge arrived. The order for this attack given directly from Hannibal to his troops, specifically designed to integrate the surrounding landscape and battle expression, demonstrates an attempt from the high command to manufacture an atmospheric condition hostile to the enemy on a battlefield.

Contrastingly, Graeco-Roman authors recorded battle expression types that produced such an overwhelming atmosphere of noise that they were worthy of mention. This atmosphere of noise was presented as being tremendous and striking in nature, which seems to have affected those that witnessed it for it has been recorded. Plutarch records the battle expression of the Ambrones tribe that consisted of united rhythmic jumping, clashing of weapons and chanting the tribal name "*Ambrones*". The noise generated by these different actions would have included the ground shaking due to the leaping of hundreds to thousands of men. The sound of massed weapons clashing, vocal noise and ground shaking created an atmosphere worthy of record.

> However, though their bodies were surfeited and weighed down with food and their spirits excited and disordered with strong wine, they did not rush on in a disorderly or frantic course, nor raise an inarticulate battle-cry, but rhythmically clashing their arms and leaping to the sound they would frequently shout out all together their tribal name Ambrones, either to encourage one another, or to terrify their enemies in advance by the declaration. The first of the Italians to go down against them were the Ligurians, and when they heard and understood what the Barbarians were shouting, they themselves shouted back the word, claiming it as their own ancestral appellation; for the Ligurians call themselves Ambrones by descent. Often, then, did the shout echo and reecho from either side before they came to close quarters; and since the hosts back of each party took up the cry by turns and strove each to outdo the other first in the magnitude of their shout, their cries roused and fired the spirit of the combatants.[5]

This Ambrones tribal cry must have been rehearsed and practiced for the performance to have been effective. The unity and clarity of noise and movement was recorded as being consistent among the participants, even when they had been gorged with food and strong wine. It is portrayed as a custom well known to the Ambrones military force. Moreover, this battle expression must have been a well-known practice within the sphere of influence of the Ambrones tribe as other tribal groups linked through kinship with the Ambrones were aware of how to perform it. This is evident in the above source where the Ligurian units, as opposed to the Ambrones in the battle, are recorded to have understood the chant and then were able to replicate it. Both Ambrones and Ligurian forces engaged in a vocal duel of the noise of the same chant that sounded as though the noise was a continuous echo across the battlefield.

Graeco-Roman military handbooks advised the high command of military forces to encourage the performance of a loud battle expression to create an atmosphere on the battlefield that would intimidate the enemy. Onosander and Vegetius refer to the military benefits of sending an army into battle utilising massed noise and visual effects to unnerve the onlooking enemy.

> One should send the army into battle shouting, and sometimes on the run, because their appearance and shouts and the clash of arms confound the hearts of the enemy. The dense bands of soldiers should spread out in the attack before coming to close quarters, often waving their swords high above their heads toward the sun. The polished spear-points and flashing swords, shining in thick array and reflecting the light of the sun, send ahead a terrible lightning-flash of war. If the enemy should also do this, it is necessary to frighten them in turn, but if not, one should frighten them first.[6]
>
> The war shout should not be begun till both armies have joined, for it is a mark of ignorance or cowardice to give it at a distance. The effect is much greater on the enemy when they find themselves struck at the same instant with the horror of the noise and the points of the weapons.[7]

Both references recommend the creation of an atmosphere where the enemy would feel comparatively small to their opponents. Onosander advised the spreading out of dense ranks of troops, while brandishing their weapons, prior to engagement with the enemy to appear more numerous and intimidating. Vegetius suggested to maximise the impact of the battle expression on the enemy by not emitting massed vocal noise until initial contact with the enemy had taken place. The discipline involved with either of these methods would have required dedicated training. From Graeco-Roman military handbooks, the military high command was influential in the preparation, through training, and execution of battle expression forms that intentionally aimed to create an intimidatory battlefield atmosphere that consisted of noise and intimidating appearance.

Graeco-Roman authors reveal that the noise produced by armies prior to violent engagement on the battlefield created a tremendous atmosphere of sound. The noise generated could be a combination of massed vocal, bodily movement, such as clashing weapons against shields, and instrumental, such as the playing of trumpets and horns. Extracts from Polybius, Livy and Ammianus Marcellinus works suggest that the atmosphere of immense noise associated with the performance of a battle expression was a feature of warfare that spanned from the 2nd century BC to AD 4th century. Polybius claimed that the combined noise of two armies performing a vocal battle expression at the same time, in unison with battlefield spectators cheering on the same forces, created a deafening atmosphere that inspired terror and acute anxiety.[8] Moreover, Polybius provides another example of a tremendous atmosphere of noise produced by a battle expression. In this instance, a Celtic military force on the battlefield undertook a vocal and instrumental battle expression that was exacerbated by their appearance to create a profoundly intimidating atmosphere.

> The Insubres and Boii drew up for battle wearing their trousers and light cloaks, but the Gaesatae had discarded these garments owing to their proud confidence in themselves, and stood naked, with nothing but their arms, in front of the whole army, . . . The Romans, however, were on the one hand encouraged by having caught the enemy between their two armies, but on the other they were terrified by the fine order of the Celtic host and the dreadful din, for there were innumerable horn-blowers and trumpeters, and, as the whole army were shouting their war cries at the same time, there was such a tumult of sound that it seemed that not only the trumpets and the soldiers but all the country round had got a voice and caught up the cry. Very terrifying too were the appearance and the gestures of the naked warriors in front, all in the prime of life, and finely built men, and all in the leading companies richly adorned with gold torques and armlets. The sight of them indeed dismayed the Romans.[9]

The physical appearance of the Celtic warriors in accompaniment with the sound of their vocal and instrumental battle expression generated a fearsome atmosphere that affected the Roman forces. Polybius records the use of innumerable horns and trumpets played in conjunction with the whole army shouting in unison creating such a noise that it seemed the battlefield joined in the performance. This description implies that the atmosphere comprising multiple senses – the sight of finely built Celtic warriors, the sound of massed voices and instruments and the feeling of the ground moving along with the noise, was overwhelming.

Livy and Ammianus record the impact such tremendous noise had on the enemy and their animals that were present on a battlefield. Previous sources described the impact the sound of the battle expression had in the creation of an intimidating atmosphere on the battlefield, the following sources describe how this atmosphere of immense noise inspired humans and animals that were confronted with this atmosphere to react. Livy claims that when a Roman army emitted a battle expression that consisted of sounding horns, trumpets and a cheer, the noise created was so tremendous that the elephants in the enemy ranks panicked and turned on their own men to flee.[10] Whether it was just the sound that made the elephants flee or a combination of noise along with missiles fired at the beasts' eyes remains uncertain. What is clear is that the battle expression created an atmosphere of overwhelming noise that animals were affected by it. Considering that war elephants were trained to be acquainted with the sounds of the battlefield, the noisy atmospheric conditions mentioned in Livy proved too much for the elephants to contend with.

Likewise, Ammianus claims that humans, too, found the noise produced in a battle expression too much to tolerate.

> and our army, aroused by the trumpets' blast, was hastening to the spot with threatening cries, the attacking force retreated in terror, though without loss.[11]

Ammianus claims that the threatening cries produced by the Roman army were the cause of the enemy retreating in terror. Whether the battle expression of trumpets and vocal noise was the main cause of the enemy's terror, and subsequent retreat, or if it was the arrival of the troops that took the enemy by surprise by the unexpectedness of the situation is unclear. What is clear is that Ammianus used the battle expression and the atmosphere it created as the catalyst for the terror inflicted among the enemy forces. Moreover, the strategical significance of this episode further provides evidence that the high command of an army was influential in the encouragement and use of the battle expression to manipulate the atmosphere on the battlefield for its own military advantage. In the extract given earlier, Ammianus claims that the trumpets were used to arouse the Roman troops into action. In the context of this episode, the arousal of the troops was required for two reasons. Firstly, the Roman force needed to hasten to the spot where the enemies were of a significant military threat. Reaching that destination on the battlefield was of importance before damage was done to the Roman interest there. It was the high command of the Roman army that issued the order for the trumpets to sound, making this feature of the battle expression, and the changing of the atmosphere on the battlefield, the responsibility of the high command. Secondly, it appears that the threatening nature of the vocal element of the battle expression was caused by the conviction of the fighting troops in the Roman army to produce such noise and atmosphere that caused terror in the enemy ranks.

However, it seems that the Roman high command expected this outcome to be achieved and intentionally used it to force the enemy to retreat from the place where the Romans were in haste to get to. Ammianus ends this extract with the statement that the enemy retreated in terror without loss, meaning that they suffered minimal, if any, casualties. The high command, it seems, aimed to force the enemy to withdraw from the place that they were threatening to with or without the use of violent engagement. The reasoning behind this idea is twofold. Firstly, the distance that the Roman force had to travel, which is not detailed in the extract, must have been quite some way considering the Romans were hastening there. So, to remove the enemy threat from that location the Romans had to either arrive at the spot quickly and force the enemy to retreat through a violent confrontation, or as the extract details, the Romans could force the enemy to retreat without bloodshed through the performance of a battle expression that altered the atmosphere of the battlefield to one that benefitted the Romans. It was the atmosphere of immense noise created by the Roman battle expression that resulted in the enemy's retreat without the shedding of blood. The Roman high command had achieved its desired outcome through the manipulation of the battlefield atmosphere.

The Germanic battle expression, *barritus*, aimed to create an atmospheric condition that could be used during battle to measure the commitment of those that undertook it. Tacitus details the purpose and objective of this cry in battle – to inspire courage among the participants through the measurement of the effectiveness of the cry. German warriors typically performed the *barritus* into their shields to gauge their individual and collective conviction for battle. Through the sound, they were able to produce and the echo from the shields German warriors could assess their unity of purpose and heart for battle.

They further record how Hercules appeared among the Germans, and on the eve of battle the natives hymn "Hercules, the first of brave men." They have also those cries by the recital of which – "barritus" is the name they use – they inspire courage; and they divine the fortunes of the coming battle from the circumstances of the cry. Intimidation or timidity depends on the intonation of the warriors; it seems to them to mean not so much unison of voices as union of hearts; the object they specially seek is a certain volume of hoarseness, a crashing roar, their shields being brought up to their lips, that the voice may swell to a fuller and deeper note by means of the echo.[12]

Tacitus claims that the warriors could determine from the volume and pitch of the sound produced from the echo into the shield how effective the cry was. The assessment recached revealed the potential of the cry to intimidate the enemy or whether the participants were timid for battle. This episode reveals the intent of Germanic warriors to create an atmosphere of tremendous sound produced vocally and extended through the echo of the cry into warriors' shields. Through the warriors raising their shields to their mouths and performing the *barritus* battle expression, the full and deep sound that was produced created an artificial terrain of noise that hills and valleys would naturally generate.

In contrast, armies from the Graeco-Roman world adopted silence as a battle expression to generate a desired atmosphere to achieve military objectives. The silent approach into battle from a military force was intentionally ordered by the military high command as a tactic to unnerve the enemy. Xenophon records the episode where Cyrus the Younger, during his attempted rebellion against his brother Artaxerxes II, advised the Greek contingents within his army to be aware of and remain resilient against the loud vocal battle expression of the Persian force. Xenophon records that this advice was misguided as Artaxerxes' force did not approach the battle with massed noise but in strict silence.[13] Xenophon adds that the Persian advance against the Greeks was not only silent but was slow so that the enemy could maintain an equal step. For Artaxerxes' force to achieve this desired outcome prior preparation before the battle would have been undertaken, as discipline would be required not to perform a vocal battle expression, particularly when it seemed to have been a traditional feature of this Persian force advancing into battle according to Cyrus' advice. The silent and disciplined atmosphere created by Artaxerxes' force clearly affected the Greek forces on the battlefield as Xenophon made note of the experience. Interestingly, it seems that this silent approach would have similarly affected the prior assumptions of Cyrus' high command who were expecting to be confronted with massed noise. It appears that the high command of Artaxerxes' force intended to surprise their opponents with their disciplined, silent approach to battle.

Experimental archaeology can be undertaken using the source material to recreate and compare the ancient phenomenon of the battle expression to other modern-day equivalents. The findings reveal that ancient military engagements of the Graeco-Roman world were highly atmospheric in massed noise, movement and/or visual illusion. This contrasts the modern-day perception of it. Modern-day practices that have similar atmospheric qualities to the ancient battle expression can be compared to the ancient evidence. Viewing recordings of crowd behaviour at sporting events during televised broadcasts of matches and from YouTube uploads, such as European football stadiums, correlate with the qualities of massed noise, movement and/or visual illusion found in the ancient battle expression. This is more evident in well-supported and organised fan bases of English and German football clubs/teams. International football competitions such as the UEFA European Championships and the FIFA World Cups are also exampling of crowd behaviour that can be compared to the performances of ancient battle expression. In a similar fashion, crowd behaviour during public protest movements, namely the chants and visual illusion of colour and banners amongst the protesters portrays, too, the qualities of the ancient battle expression. The crowd behaviour in football stadiums and on the streets during protest movements is highly organised in the performance of chants/songs and the attire worn by crowd members. The songs/chants coupled with the attire worn by the supporters of football teams and the protesters suggest effective levels of pre-planning, rehearsal and organisation from a leading group within. The sophistication of rehearsed, as well as spontaneous outbreak of song/chant/movement/attire, is reflective of the efforts made by ancient military forces in the Graeco-Roman world to intimidate the enemy and inspire the group participants.

The use of modern-day technologies results in the ancient literature finding *life*, such as recreating confirmed battle expression (from ancient literature) using computer-based audio recording technology. From this, an accurate sound can be generated using the voices of thousands of men which reveals the impact an ancient battle expression from the Graeco-Roman world had on the enemy and its participants.

This study has the potential to recreate the sound of a battle expression used in the ancient world. The ability to recreate a battle expression permits an understanding of the impact it had on a participant in, or witness to, a specific battle. Confirmed battle expression types, as identified through literary and/or archaeological sources, may be reconstructed using modern voices and computer technology into a recreation of the original battle expression. The computer program "Audacity"[14] allows for one, or multiple voices to be recorded and then, using the various attributes of the program, one voice can be multiplied into thousands of voices, generally through copying and pasting existing recordings onto the same file. The other voices can be altered to create a more authentic "crowd" noise by delaying the cries of certain voices and changing the pitch and tone of other voices to accommodate for individual voice differences and delays along a front-line fighting force in antiquity. Other special effects such as adding an "echo" can be applied onto battle expression types that were known to have been used within geographical environments that reflected the sound of the noise off hills, mountains or in valleys.[15] To provide viable *comparanda* extracts from the primary sources will be compared to the sound generated by European football supporters inside stadiums and in other purpose-specific contexts. This process creates a link between the modern atmosphere and sound generated by European football supporters to the sound made by the armies from the ancient Mediterranean world before, during or after battle.

The incorporation of traditional literary-based research of ancient texts to gather information to generate an accurate sound, and impression, of an ancient battle expression using other mediums, namely audio recording technology, is innovative. This approach can reveal the impact battle expression types had on the enemy and the military force undertaking it. Significant, interrelated methodological problems arising from the limitations of the ancient literary record; the paucity of modern scholarship relating to the topic and, importantly, the theoretical complications involved in identifying, measuring and comparing an aural–oral phenomenon extracted from narrative sources with a quantifiable modern equivalent can be overcome using experimental archaeology. Creating a link between the modern ambience generated by European football supporters and the sound made by the armies of the ancient Mediterranean before, during and after the battle, may demonstrate how a traditional philological approach to data collection can be used in conjunction with other applications (e.g. sociolinguistic analysis, statistical sampling, audio recording technology) to generate an accurate sound-impression of an ancient battle expression.

Notes

1 Fron. Strat. 2.4.3.
2 Arr. Ana. 1.6.1–4.
3 Amm. 19.2.11–12.
4 Livy. 22.4.6–7.
5 Plut. Mar. 19.3–5.
6 Ono. Strat. 29.
7 Veg. DRM. 3.18.
8 Polyb. 18.25.1–2.
9 Polyb. 2.28–29.
10 Livy. 30.33.13.
11 Amm. 24.5.9.
12 Tac. Ger. 3.
13 Xen. Ana. 1.8.11.
14 Audacity 1.2.6 'A Free Digital Audio Editor' Build date: Nov 13 2006. Website: http://audacity.sourceforge.net/.
15 This effect could be applied, specifically, for the recreation of the "din" generated by Alexander the Great's army on the Danube. Here, Alexander ordered his army to raise their "war cry" and to clash their weapons together against their shields, so that the sound generated from the army and the reflection of nearby mountains put the enemy to flight. Arr. Ana. 1.6ff.

4 Group cohesion

The battle expression paradigm captures the complexity of human experience and psychology in warfare. It is important, therefore, when formulating a complete understanding of this phenomenon, to consider not only those actions that take place *during* battle but also the various types of expression prior to the engagement of conflict that could be undertaken in the moments leading up to the battle. Our ability to identify categories of expression can help us to shed light on the range of needs informing the performance of those expressions whether they be the needs of the state, the community of combatants, or of individual soldiers. In this context, we may speak of broader sociopolitical imperatives such as patriotism or imperialism, or similarly encompassing sociopolitical drives like glory or honour. What seems certain is that the heterogeneous backgrounds and psychology of men fighting in a military force inspired, informed and framed what happened before as much as during violent conflict.

The search for group support and belonging to feel protected, for example, was an instinctual need of men before battle. The methods that men used before and during battle to measure the psychological and military strength of their fighting force came in various forms. Often these methods were expressed *en masse*: singing, praying, imitating wild animals, the playing of instruments, gesticulations, taunting, appearance and marching. The more cohesive battle expression often resulted in greater enthusiasm developed amongst the fighting ranks. Of note, the more enthusiastic an ancient military force appeared or sounded before battle did not mean that they would be the superior military force that gained victory. This argument means that a military force undertook a battle expression for the purpose of generating conviction and cohesion within the ranks. Additionally, public demonstrations of a battle expression served to gauge the mental state of the enemy who bore witness to it. The potential for a battle expression to influence the outcome of a battle, by denting the enemy's morale thereby weakening their military effectiveness or preventing bloodshed through forcing the enemy to surrender or withdraw from the battlefield, reveals its military significance. Literary sources suggest that the battle expression was encouraged by the military high command due to its effectiveness to inspire an army and intimidate the enemy.[1] This chapter will focus on the evidence that battle expression was utilised by ancient military forces to gauge and strengthen unit cohesiveness for battle.

DOI: 10.4324/9781003280439-4

The psychological dimension of the battle expression

Modern research into the psychological dimension of battle has gained momentum over the last three decades. While much research focuses on modern combat experiences and conditions such as PTSD/CIS, some work has compared modern and ancient combat experiences, while others have focused on the nature of ancient combat. Each type of study gives valuable insight into the reasons why ancient military forces undertook battle expression prior to battle. These reasons include – to remind men of their training to bring a sense of familiarity to the battlefield, building unit cohesion for greater efficacy and gauging conviction for battle by the high command. While helpful in these areas, modern studies about the psychological dimension of battle fall short in completely understanding this practice, fortunately, ancient literary sources help overcome these shortfalls on the purposes behind these performances.

Studies into the motivational and behavioural aspects of the modern soldier in battle can be applied to the ancient Graeco-Roman battle expression.[2] The use of modern conflicts as case studies to demonstrate this does not limit its application to ancient warfare.[3] It should be reasonably clear, for instance, that at the core of any soldier's mental attitude prior to the engagement of battle is his combat preparedness. For instance, military training is fundamental for an ancient or modern soldier's practical and psychological preparation for battle.[4] The ability for a soldier to become acquainted with the sights and sounds of battle is integral in creating confidence and preventing the outbreak of fear.[5] Ancient forces adopted the battle expression to assure that the mindset of the combatants would default to the patterns established with familiarised exposure during their training, and diffuse any problematic concerns associated with the uncertainty and confusion[6] that the battle experience could bring to inexperienced troops.[7] Through the promotion of unit cohesion, individual members would become more devoted to the group and, therefore, more enthusiastic to honour the unit in battle.[8] This provides a useful explanation that can be applied to the numerous instances in the surviving literary record of ancient armies expressing their identity, in their various forms, on the battlefield.[9] Group cohesion on the battlefield generated a sense of reciprocal obligation to protect all members of that group.[10] The battle expression is therefore a helpful heuristic frame *and* military practice by which to identify different groups on the battlefield and to gauge their cohesive strength through its performance.

Another helpful modern perspective to consider in this regard is the rhetoric of Graeco-Roman combat. For example, the Greeks had great respect for the contrasting and rapidly changing moods of soldiers in battle.[11] Xenophon respected the influence psychological factors had on soldiers in battle:

> It is neither numbers nor strength which wins victories in war; but whichever of the two sides it be whose troops, by the blessing of the gods, advance to the attack with stouter hearts, against those troops their adversaries generally refuse to stand. And in my own experience, gentlemen, I have observed this other fact, that those who are anxious in war to save their lives in any way

they can, are the very men who usually meet with a base and shameful death; while those who have recognized that death is the common and inevitable portion of all mankind and therefore strive to meet death nobly, are precisely those who are somehow more likely to reach old age and who enjoy a happier existence while they do live.[12]

Xenophon claimed that psychology was the most important factor in winning battles.[13] High levels of visual confidence in one's army could be enough to produce panic in the enemy whilst signs of fear could raise the spirits of the enemy.[14] For Xenophon, the gods could inspire confidence within an army and, just as easily, spread fear and panic.[15] Panic was viewed in Greek warfare as being a contagious effect that could not be easily reversed.[16]

Phalanx warfare in the classical Greek world consisted of densely packed units of hoplites. The hoplites' mentality was affected dramatically on the battlefield prior to engagement with the enemy.[17] The time spent facing the enemy from a distance, the knowledge of, or lack thereof, regarding the types of wounds that could potentially be sustained, and the experience of being hard pressed all around by friend and foe that resulted from phalanx warfare affected the hoplite's psyche.[18] Hoplites would develop feelings of fear and trepidation for the battle that was to come. Greek phalanx warfare was by an agreement whereby opposing Greek military forces would arrange to meet face-to-face on a given day. Military forces would eye each other off anywhere from a few minutes to hours on end. During this time, Greek phalanx armies would attempt to intimidate the other to reverse the onset of fear amongst friendly troops and to accelerate the levels of anxiety within the enemy. Military forces that were successful at this could potentially win the battle before engagement with the enemy, as the latter would break and flee. Greek armies institutionalised ritualistic battle expression, such as the paean to achieve this.[19] Aside from the *paean*, the Athenians raised a collective utterance of *eleleleu*[20] on the battlefield before advancing on the enemy. The yelling and singing of the men in the phalanx pre-battle aimed to rouse the men for battle. Therefore, yelling and singing generated a sense of regimental spirit amongst the troops and was one reason why Greek hoplites resolved themselves to fight in this brutal manner of warfare. The use of song and symbolism that connected the rank and file along sociopolitical lines inspired different Greek battle expressions, so too, did the commander of the phalanx who was used as inspiration for hoplites in the lead up to the battle.[21]

Evidence that ritual song and vocal noise before battle resulted in psychological benefits for a Greek army (heightened levels of regimental spirit and collectivism) correlates with the battle expression concept. The limitations of the term war cry, when attempting to account for a military phenomenon, such as massed vocal noise before engagement with the enemy, is demonstrated as the *paean* and Athenian *eleleleu* does not categorise the purposes associated with such actions to war cry. Instead, the study of the nature of phalanx warfare helps to explain why massed vocal noise was undertaken on the battlefield: to instil resolve within the individual and group for the manner of fighting that would ensue, or to potentially prevent it from occurring. Despite revealing insight into the battle expression concept, the nature of phalanx warfare fails to acknowledge the social, political and religious significance that types of battle expression may have had for a military force. A battle expression reflected group identity, often based on ethnicity, patriotism or political ideology. Social customs and religious beliefs were revealed through the different types of battle expressions. The political disunity within ancient Greece, evident through the *poleis*, highlighted social and political diversity that were reflected in battlefield customs that are not accounted for when studying the nature of phalanx warfare.

The collective psychological state of hoplites in a phalanx was significant for its effectiveness on the battlefield.[22] The depth of a phalanx formation gave the hoplites a psychological lift and exerted a psychological pressure on the enemy with the sight of an unbreakable phalanx. The Spartans created a psychological advantage for themselves in battle as they did not run and shout into battle to keep up their courage but remained calm, measured and quiet as they advanced to the sound of the flute.[23] Psychological support that being in a phalanx army would have provided a hoplite and the motivation that army would have gained from undertaking massed vocal performances, such as the paean, prior to battle.[24] Ancient literary evidence suggests Greek hoplite armies did suffer from fear and trepidation in the lead up to the battle.[25] Undertaking battle expression aimed to unite the phalanx and instil the hoplites with calm before the violent confrontation with the enemy.[26]

Recognition of psychological imperatives was not limited to Greek writers, *strategoi*, or combatants; so too, quintessentially Roman military men like Caesar expressed their awareness of such drives, and of how best to apply the lessons learned to the exigencies of battle, whether in training, as part of the customary patterns of activity prior to engagement, or on the field.[27] The *animus* of soldiers, for Caesar, is easily alarmed by any surprise or anything unfamiliar.[28] Inexperienced soldiers are especially vulnerable to panic, while soldiers who were experienced with the battlefield environment grew immune to fearful situations.[29] Caesar's generalship suggests that he always attended closely to the relative *animus* of his own and the enemy's army since it governed their fighting quality. When Caesar refused battle, he hoped that the psychological dominance of his army would be so great that the enemy would surrender without any bloodshed.[30] In the lead up to the battle, the *animus* prescribed suggestions to the general to deploy and manoeuvre for psychological reasons, to hurt the *animus* of his foes and to increase that of his own soldiers.[31] This is demonstrated in Caesar's recommendation to allow an army to undertake a battle expression and his subsequent criticism of Pompey at Pharsalus for not permitting his army to do so.[32]

Modern research's exposure of the psychological dimension of warfare contributes greatly to the study of the battle expression. Coupled with Graeco-Roman literary source material,[33] this perspective suggests that those belonging to the highest echelons of command in the Greek and Roman military influenced, shaped and embedded the battle expression in state-sanctioned codes of doctrine and training. This ensured that those facets of expression deemed integral to military success were designed for mass participation to maximise effective communication, acquisition and application. Caesar's claims are at the heart of this. The vested interest the military commanders had in encouraging their army to partake in a range of battle expressions served multiple purposes. Initially, it helped the high command gauge the level of commitment for battle from the enemy as well as their own army. From this, the high command could devise and achieve military objectives. The high command promoted the use of intimidating acts on the battlefield as a tactic to avoid bloodshed and achieve victory over an enemy. Battle expression forms were encouraged by the high command to strengthen the resolve and instil camaraderie within their army. Military commanders would not have begrudged their men the means to become motivated through socio-cultural beliefs and practices that inspired men to endanger their lives.[34]

The notion that the high command was instrumental in promoting the battle expression is evident in Onosander's military handbook, where a section is dedicated to the recommendation of sending men into battle shouting:

> One should send the army into battle shouting, and sometimes on the run, because their appearance and shouts and the clash of arms confound the hearts of the enemy. The dense bands of soldiers should spread out in the attack before coming to close quarters, often waving their swords high above their heads toward the sun. The polished spear-points and flashing swords, shining in thick array and reflecting the light of the sun, send ahead a terrible lightning-flash of war. If the enemy should also do this, it is necessary to frighten them in turn, but if not, one should frighten them first.[35]

Graeco-Roman military handbooks produced by Vegetius and Maurice reveal the integral role the military high command had in relation to the integration of battle expression within the army.[36] Within the Roman army, nothing in relation to battlefield custom appears to have been undertaken without the high command's knowledge and support. Josephus claims that nothing in the Roman army is done without a word of command.[37] Ammianus, in his referral to a decline in Roman military discipline, suggests that the practice of battle expression types was commonplace in military training:

> To these conditions, shameful as they were, were added serious defects in military discipline. In place of the war song the soldiers practiced effeminate ditties.[38]

Tacitus' reference to Percennius, a ring-leader of a military mutiny at the end of Augustus' reign and the dawn of Tiberius', may suggest the intentional recruitment and use of specialised initiators of battle expression within the fighting ranks of the Roman army:

> In the camp there was a man by the name of Percennius, in his early days the leader of a claque at the theatres, then a private soldier with an abusive

tongue, whose experience of stage rivalries had taught him the art of inflaming an audience.[39]

In the context of ridiculing Junius Blaesus (commander of three legions stationed together in summer quarters) for the degeneration of discipline amongst the troops, Tacitus may unintentionally highlight a key element of the Roman battle expression. Theatrical claque leaders had experience in raising the levels of enthusiasm of a crowd in support, or opposition, of a pantomime actor.[40] Perhaps these qualities were utilised in the Roman army to instigate traditional and spontaneous forms of battle expression to galvanise the fighting ranks of the troops that they were embedded. Given the association theatrical claque leaders had "with an abusive tongue", this would have been suited for initiating taunts on the enemy. The claque leader's skill in seizing the moment – by exploiting a crowded and volatile atmosphere (on the battlefield) using satirical wit – may shed further understanding on the *revocare* episode mentioned later in this chapter.[41]

The tradition of the Roman army taunting and stimulating an atmosphere using massed vocal demonstration is evident in the military triumph.[42] The organisation and effectiveness of the ritual singing and chanting undertaken by Roman soldiers, as they marched through the streets of Rome celebrating, often at the expense of their commanding general's reputation, reveals a certain level of leadership and training within the fighting ranks. The sphere of Roman military life – in training, on the battlefield and during triumphal processions – suggests that there was a need for the services and attributes of theatre claque leaders. These men were adept in the formulation of relevant material for taunting, they could exploit specific situations and knew how to prompt the men into vocal action. Suetonius and Tacitus reveal a clear military association with theatre claques during Nero's reign.[43] These references suggest that theatrical claque leaders were known to the military high command (the emperor and his associates) and that soldiers were acquainted with crowd protocol in the theatre. Maurice's *Stategikon* suggests that the Roman army had an officer class of singers.[44] Translated from the Latin word *cantator* to mean an officer class called "Heralds", the word should be understood as an arch chanter or a lead singer. If this is the case, the Roman high command intentionally ensured the effective undertaking of vocal battle expression types through the employment of a specialised officer class, perhaps with the skillset of a theatre claque leader such as Percennius.

The theory that the military high command of armies from the Graeco-Roman Mediterranean world endorsed and saw benefit in the use of battle expression types is supported through literary references that demonstrate the systematic nature of their initiation on the battlefield.[45] Ancient literary sources explicitly state that the battle expression was ordered directly by the military high command and customarily initiated the start of the battle.[46] A clear example of this is evident at the battle of Gaugamela where Alexander the Great ordered his army to maintain strict discipline and cohesion during the battle so that their small numbers, in comparison to Darius' larger army, may still have an effectual impact on the battle, particularly by way of battle expression.

To keep perfect silence when that was necessary in the advance, and by contrast to give a ringing shout when it was right to shout, and a howl to inspire the greatest terror when the moment came to howl; they themselves were to obey orders sharply and to pass them on sharply to their regiments, and every man should recall that neglect of his own duty brought the whole cause into common danger, while energetic attention to it contributed to the common success.[47]

The ancient literary source material presents the battle expression as a means for military forces to instil group cohesion and resoluteness within their fighting ranks for the battle at hand. It was customary for different military forces to undertake similar types of practices in the lead up to or during battle to strengthen the psychological resolve of the troops. Pre-battle customs performed by large numbers of fighting men testify to the significance these customs held in focusing on the emotional state of those about to commit violent confrontation. Ancient authors refer to a diverse array of actions that many cultural groups adopted for this purpose including long bouts of singing, drinking and/or choreographed movement such as dancing in the hours leading up to the battle. Often, ancient authors did not acknowledge the cultural meaning of these practices, instead merely referring to their existence. Whether these military practices held socio-religious significance is not communicated. These cross-cultural practices were communal in nature, suggesting they were familiar and inclusive actions that catered for many participants. The pre-battle timeframe battle expressions were performed revealing their purpose to prepare those who were about to fight. The authors portray these battle expression types with the primary intention of inspiring the participants rather than intimidating the enemy.

For example, Herodotus refers to a battle expression performed by a Lydian military force during the reign of king Alyattes (ca. 619–560 BC). This Lydian force had invaded the lands of Miletus to besiege the city. Herodotus claims that Alyattes ordered his army to invade Miletus, marching to the sound of pipes, stringed instruments, bass and treble flutes.[48] Herodotus may have used this episode to stress how easy and un-resisted Lydian raids were.[49] This reference does not specify whether the musicians playing these instruments were members of the military force or musicians who accompanied the troops out of Lydian land. It is likely that the Lydians did have musicians within the ranks of the army, just as the Greek and Roman forces contained flute players and trumpeters. The practicality of flutes and pipe instruments being used during battle is probable. However, the reference to stringed instruments in this account suggests that at least these musicians were ceremonial courtiers rather than military personnel and did not partake in military affairs, such as a siege. The acknowledgement of these stringed musicians highlights the difference in Lydian military culture when compared to other ancient Mediterranean cultures, such as that of Africa, Greece and Rome. The opulence of a combined retinue of musicians that consisted of not only pipes and a diversity of flutes, but stringed instruments is distinctive of Lydian custom. From Herodotus, the ancient Lydians used musical instruments, and undoubtedly common tunes, for military affairs. The invasion of a foreign land and possibly the entry into enemy territory, where the battle could occur at any stage, resulted in the Lydians employing the services of musicians to inspire and unite their military along cultural lines.

It is evident that the battle expression of African military forces recorded in the surviving annals of historical narrative followed a scope and sequence which were consistent amongst other cultural groups across the Graeco-Roman world.[50] For instance, Polybius refers in his *Histories* to two separate Carthaginian military forces that praise their military leader in an attempt to initiate engagement with the enemy. Each military force did this through united loud vocal noise. The first example relates to the popular feeling and trust amongst the Carthaginian army displayed towards their Spartan leader Xanthippus.[51] Due to the Carthaginian army's training and organisation under Xanthippus' leadership in the course of preparations for battle against the Romans, Polybius claims that the Carthaginian soldiers, within specific groups, sporadically shouted out *en masse* the name of Xanthippus. The purpose of this loud massed yet coherent noise was to encourage Xanthippus to lead the Carthaginian army into battle against the Romans, who were by this stage confident in their military training and organisation, effectively, battle-ready because of him. The interesting feature about this reference is that Polybius specifically refers to the Carthaginians crying out in groups the name of Xanthippus, which could suggest two things. Firstly, these groups could have been formed along different ethnic or cultural lines within a Carthaginian army that consisted of a variety of cultures and tribes. This could possibly suggest that prior planning went into the performance of this battle expression amongst the different contingents of a multinational force and when they were going to exclaim the name of their popular military leader. Alternatively, these groups could suggest a spontaneous act, where the warriors within this Carthaginian army shouted out in groups from different parts of the camp after they were inspired to do so, after a group initiated the chant. The second reference relates to the outbreak of a loud coherent shout from a Carthaginian army for their leader, Himilco, to lead them into battle.[52] Polybius claims that this shout occurred after Himilco gave a motivational speech to the army as they were being besieged by a Roman army. Despite the fact the enemy probably could not hear this battle expression, it must still be regarded as one due to the motivation it instilled within the troops it was addressed to. Polybius claims that after Himilco gave his speech, the troops applauded him as one and there were loud shouts for him to lead them out against the enemy.

Sallust's history of Rome's war against Jugurtha's Numidian forces contains references to different types of Numidian battle expressions that aimed to strengthen the resolve of friendly troops. Of interest, the foreign nature of the Numidian battle expression, from a Roman perspective, is made evident in Sallust's account of the siege at Zama ca. 109 BC. According to Sallust, two separate military engagements took place during the siege. The first military engagement involved Roman military units attacking the walls of Zama itself and the Numidian forces inside attempting to repel them.[53] The second engagement took place around the Roman military camp situated on the plains outside the walls of Zama.[54] Sallust claims that Jugurtha led Numidian cavalry from Zama to attack the Roman camp outside in an attempt to force the Romans to withdraw from the walls of Zama. It was during this stage of the battle that Sallust refers to a Numidian battle expression, which correlates to other references later in Sallust's history of this war.[55]

The Roman general, Marius, while leading the Roman attack against the walls of Zama, noticed that whenever there was a break in fighting on the walls, the Numidian forces stationed there would eagerly turn to watch, in the distance, the fighting that was taking place around the Roman camp where Jugurtha had led his cavalry. It was during one of these lulls in fighting on the walls that Sallust claims the Numidians broke out into a battle expression aimed at encouraging the far-off Numidian cavalry.[56] The purpose of this Numidian battle expression was to motivate fellow Numidian warriors on the other side of the battlefield. This suggests the Numidians stationed on the walls could be seen and heard by the other Numidian force. This notion is in stark contrast to the modern under-standing of a war cry which, generally, gives the impression that military forces aimed to encourage fellow warriors in a common battle-line or to strike fear into the enemy force directly opposing them. This instance goes beyond the tradi-tional conceptual framework. According to Sallust, the nature of this Numidian battle expression involved Numidian warriors shouting and moving their bodies in what seemed to be a common fashion.[57] The expression comprised of shouts, verbalising words of warning or encouragement, along with hand gestures and the swaying of bodies, as if the defenders of Zama were attempting to avoid fire from darts.[58] Frustratingly, Sallust does not mention anything more regarding this particular battle expression:

> But whenever the besiegers relaxed their assault ever so little, the defenders of the walls became intent spectators of the cavalry battle. As Jugurtha's for-tunes shifted, you might have seen them now joyful, now alarmed; and acting as if their countrymen could see or hear them, some shouted warnings, others shouted encouragement or gesticulated with their hands or strained with their bodies, moving both this way and that as if dodging or hurling weapons.[59]

Clearly, Sallust recorded this episode due to its peculiarity from a Roman perspec-tive. The idiosyncrasies of the expression might explain why the historian does not qualify the meaning or purpose – whether religious, cultural or strategic – explaining the gesticulations or body movements performed *en masse* by Numid-ian soldiers stationed on the walls at Zama. When this battle expression is viewed from a Numidian perspective, it is evident that the gesticulation or hand move-ments made by the warriors on the wall would be recognisable to the Numidian cavalry in the field outside the walls. Even though it is not known what form or appearance these hand gestures or gesticulations took, it is evident that the Numidians were privy to their appearance and meaning, otherwise, hundreds to thousands of warriors would not have made them to communicate to a similar aggregation of men engaged on the other side of a battlefield. Interestingly, this battle expression appears to be unique to the Numidians. While Sallust does not mention the appearance of the gesticulations, he does compare the body movements to that of men trying to avoid being struck with darts hurled by the attacking force. This would logically imply a good deal of ducking, weaving from side to side and jumping. The common understanding among the Numidians regarding the combination of gesticulations and body movements enacted on the battlefield – or, in this instance, related arenas of conflict across a battle environment – further suggests that these were culturally specific and may well have been performed and practiced by Numidian warriors. The precise nature of the relationship between this composite battle expression and Numidian culture must remain highly speculative.

Sallust's account of the siege of Zama provides additional information regarding another type of Numidian battle expression that aimed to unite the military force the night before the battle. In the preamble to a military engagement, a Numidian force, consisting of Numidians, Mauretanians and Gaetulans allied to Jugurtha, had surrounded a Roman force on top of two hills. As day turned to night, the African force, having surrounded the Romans, began shouting and singing through the night in anticipation of the next day's battle.[60] This reference, like the earlier narrative episode, provides some insight into the spectrum of battle expression associated with African warriors. Sallust claims this practice was customary of these "barbarians",[61] emphasising the otherness or cultural *difference* of foreign battle expression compared to Roman. From an African perspective, the tradition of making loud noise during the night before battle must have been significant and appears to have been a normal feature prior to battle. The songs, hymns, shouts or dances that would have been performed are not mentioned, however, the subject of the songs must have had culturally familiar or resonant religious, mythological, historical or social undertone for the majority, if not all, in the composite African force to participate. According to Sallust, the battle expression of specifically Numidian and broader African military forces, contrary to the Roman perspective, were both geographically and ethnically contextualised *and* sophisticated, suggesting a strong cultural connection or familiarity amongst most of the participants to the battle expression performed.

Plutarch's *Life of Sulla*[62] details the battle expression characteristics of the republic and early imperial military forces as being culturally Roman. In this extract, Plutarch describes the battle of Chaeronea from ca. 86 BC. In the initial stages of this battle, the chariots of Mithridates' Asian force had launched an attack on the Roman front line. The attack against the Roman line had little effect due to the chariots not having enough open space to build up momentum for their attack. The chariots had made little impact on the Roman force opposing them and Plutarch claims that their attack was quite sluggish and feeble. The Romans, who had easily beat off the chariot attack, spontaneously let out a battle expression that was inspired by the recent events.

Plutarch claims that the Romans *en masse* applauded the chariots, sarcastically, and laughed at their poor attempt to break through their ranks. The Romans then began to shout out to the enemy together, no doubt inspired by one or two individuals who initiated the taunt, to start again or have another go. The interesting aspect to this taunt is that Plutarch adds that the call made by the Roman soldiers to their enemy to start over or bring on some more chariots was the same call spectators in the circus would shout out *en masse* during the Roman chariot races.[63]

In Latin, the language presumably spoken by the Roman soldiers, the verb *revocare* means to recall. In Ovid's work, *Amores*,[64] he states that during a race he was watching with his, would-be, girlfriend, the crowd shouted out *revocare* to re-start the race after a poor start impacted on the quality of the race. This was a common occurrence at Roman circus meets.[65] Therefore, Plutarch, who wrote in Greek but was familiar with both popular Roman customs and idiomatic expressions in Latin, is most likely alluding to Roman humour being exhibited in this Roman battle expression. The soldier's spontaneous taunt during this battle expressed sarcasm towards the enemy for their chariots to begin their attack afresh:

> For these are of most avail after a long course, which gives them velocity and impetus for breaking through an opposing line, but short starts are ineffectual and feeble, as in the case of missiles which do not get full propulsion. And this proved true now in the case of the Barbarians. The first of their chariots were driven along feebly and engaged sluggishly, so that the Romans, after repulsing them, clapped their hands and laughed and called for more, as they are wont to do at the races in the circus.[66]

The battle expression detailed in Plutarch's work is clearly sophisticated in nature, through united and spontaneous clapping and chanting, and purpose, which was to denote Roman military superiority, uniting the Roman troops together against an enemy force and lifting morale. The massed vocal battle expression at Chaeronea served to reaffirm the soldiers' identity against a non-Roman enemy.

Roman military practice likewise incorporated battle expression that increased the cohesiveness and effectiveness of the army. Caesar in his work *Civil War*[67] explains, as a side note to his historical account of the battle of Pharsalus, that Pompey's tactic to order his troops to be silent and still prior to engagement with the enemy was flawed. Caesar claims in this passage that military generals should encourage the means to extract spirit and keenness for battle amongst their troops – that is through allowing and encouraging battle expression to be used. Caesar goes on to say that raising a battle expression not only puts both fear into the enemy's hearts but also instils motivation within the men associated with the cry. In this same extract, Caesar acknowledges that the practice of war cries was ancient in his day and had been in use ever since for sound reasons. This suggests that it was uncommon for Roman armies not to initiate a battle expression before, during or after military engagements. Therefore, Caesar claims that performing a battle expression was a typical feature of military life that had been a continuing facet of military practice since earlier times, or former times to his day, which he states as: "*antiquitus institutum est*".

Similarly, a feature of Celtic/Germanic battle expression was to motivate the warriors before battle through the involvement of the whole community gathered at a battle. Tacitus and Caesar refer to battles in which the Celt/German army had brought with them their womenfolk to the battle.[68] According to these sources, women were brought to the battlefield to inspire the males to fight more vigorously to prevent the symbol of the family (women), from falling into the hands of the enemy should the men be unsuccessful in battle. The presence and employment of women and children (the family) in Celtic/Germanic battle expression[69] is a unique feature rarely seen in any other ancient culture within the Graeco-Roman world. Caesar details the nature of a German battle expression that included the womenfolk. As the Romans formed up on the battlefield, the Germanic army along with their women cut off any hope of retreat for the German warriors by positioning their wagons at their rear.[70] As the German males formed on the battlefield, the German women implored the men, through shrieks, dishevelled looks and tears, not to allow them to fall into Roman slavery. The creation of a physical barrier and the motivation of the women highlights how important German/Celtic society viewed the natural world – which was, in turn, featured on the battlefield. The womenfolk, the symbol of family, and the barriers made of wagons, replicating a natural obstacle difficult to penetrate were used to motivate the German warriors to defend their families. The impact on the Romans, too, would have been to highlight the determination of the Germans to fight more stubbornly.

In another instance, Tacitus and Livy highlight culturally unique characteristics of Celtic/Germanic battle expressions that psychologically readied the men for battle. Both authors refer to the Celts/Germans as performing actions and/or songs that were solely associated with their culture. Livy describes the actions and songs performed by the Celts before battle as following some ancestral custom – *in patrium quendam modum*.[71] The songs that the Celts sung, the way in which they displayed themselves, such as their long flowing hair, tall physique, large shields and weapons, and the actions they adopted, such as leaping into the air, battering their shields together and howling like animals, was to Livy, a traditional practice that entailed deep meaning and significance. For Livy to make note of this suggests an element of respect and sophistication in the processes undertaken by the Celtic warriors, despite the alien nature of it. Tacitus refers to the songs German military forces sang in the lead up to the battle. According to Tacitus, Hercules/Herakles was a hero that visited many peoples and places, however, it was the Germans who primarily sang songs about him before the battle.[72] Whether Tacitus is accurate with the name of this hero, or whether it is a hero or God, such as Wodin or Beowulf (or other) with similar traits as Hercules remains to be seen. What is clear is that the Germans before battle sang to and about heroes of their mythological and/or historical past. Battle expression types were not based on primitive noise and lack of coordination, but on the contrary, were spiritually meaningful presentations that aimed to culturally unite the military force. This feature consistently appears within Graeco-Roman literary works and is common among the cultures from the Graeco-Roman Mediterranean world.

The Persian army of AD 4th century adopted battle expressions that gauged the cohesiveness of their troops. Persian insurrection into Rome's eastern borders under Sapor II resulted in the Persians, and their allies, besieging the Roman-held city of Amida in Mesopotamia in AD 359. Ammianus describes a variety of battle expression types undertaken by the Persian army during this siege that prepared them for battle. The initial stages of the siege revolved around the death of a notable Persian prince, the son of Grumbates king of Chionitae and ally of the Persian king Sapor. According to Ammianus, Grumbates was sent forward, by Sapor, under the walls of Amida to negotiate the terms of surrender for the city. As Grumbates came forward, a Roman marksman stationed on top of the walls of Amida fired a shot at Grumbates and hit his son next to him. The bolt fired pierced the armour of the prince who died. Several tribes within the Persian army became so outraged at this loss (no doubt the tribes from the kingdom where this prince originated) that it roused these tribes into instigating a battle expression. The battle expression consisted of harsh cries and was unique to them.[73] The use of harsh cries to depict this Asian battle expression suggests that the theme was full of anger and hate which would have reflected the emotions the tribes within the Persian army felt after the death of their prince. Ammianus most probably did not understand the lyrics or recognise the tune to the unified massed noise which was performed in native dialects. Despite not knowing the actual wording of the battle expression, the context of it being initiated suggests the tribes aimed to invoke vengeful forces known to the Asian tribes to be set upon the enemy and/or to install pride and honour into those still living and able to seek retribution for the fallen. Shortly after the Persian army initiated a full-frontal attack on the defences of Amida by the entire Persian force. Ammianus describes that the entire Persian army clashed their weapons together, to generate an enormous sound, as they charged at the city walls.[74] Ammianus does not detail the length of time the Persians clashed their weapons for or the pace to which it was set. The act of clashing weapons together reveals that the noise generated by thousands of men would have lasted quite a considerable length of time. Whether this battle expression was traditional or spontaneous, the participation in such a performance would have encouraged the participants to maintain a certain level of noise that would have been morale-boosting. The impact of tens of thousands of Persian soldiers, located at different areas around a large settlement, spontaneously or purposefully clashing their weapons together[75] would have generated an awesome atmosphere that would have affected both sides quite differently.

Spontaneity of the battle expression

Battle expression types could be both pre-planned and spontaneous in their creation. The events that transpired during a battle could, at times, inspire the undertaking of a battle expression to strengthen the resolve of a fighting force or capitalise upon the enemy's misfortune to further boost unit morale. The act of exploiting given circumstances on the battlefield that could expose the enemy for ridicule, or serve to motivate a military force, was yet another way of affecting the psychology of men in battle. As will be explored subsequently, spontaneous taunting of the enemy was a common feature of a battle expression. However, other forms of spontaneous expression could develop as events on the battlefield unfolded.

Xenophon demonstrates that military forces could choose from a spectrum of possible expressions during the initial stages of a battle that aimed to enthuse and unite the army that performed it. For example, Xenophon describes the actions of Cyrus in the opening stages of a battle fought between Persian and Assyrian forces during the king's invasion of Assyria.[76] After the traditional pre-planned battle expression was performed, Xenophon refers to the formulation of two spontaneous types of battle expression that both began as a result of individual cries and resulted in the majority of Cyrus' army taking part. The first of these spontaneous battle expressions broke out immediately after the completion of the traditional paean, where individual soldiers within the front ranks of Cyrus' army cried out to neighbouring soldiers' words of encouragement. The individual cries led to many others taking up the same cry until a battle expression developed whereby different sections of the army communicated to each other through chanting the words, initially instigated by one or more individuals. According to Xenophon, the frontline ranks, united in voice, urged the rear ranks to move forward bravely with them. In response, the rear ranks, united in voice, chanted for the forward ranks to lead them in doing so.[77]

This type of battle expression from the ancient world has parallels in the modern day within European football stadiums. Amongst common supporter groups, during a football match, who are seated in different sections around a stadium; one section, or grandstand, prompts the other sections to respond through chants. A famous version of this practice derives from England where supporters identify themselves to each other based on that *side* of the opposing team's supporters they are located, opposing supporters are usually segregated to one small section of an *away* stadium.[78] Another version of this type of practice originates from Serbian football supporters who, when the national team is playing in a stadium, cry out the word "*Serbia*" in a sequence around the ground that is comparable to the process used to undertake the popular Mexican Wave. Xenophon claims the effect that traditional and spontaneous battle expression types had on the Persian army, during the preliminary stages of this battle, was substantially positive due to the army's buoyant attitude.

The second spontaneous battle expression broke out as soon as the Persian army came within ballistic range of the enemy. Xenophon claims that Cyrus himself initiated this battle expression by shouting out repeatedly to those around him as he charged towards the enemy "Who will follow? Who is brave?".[79] The result of this, according to Xenophon, was that the whole of the Persian army took up the same shout together as they charged towards the enemy. This reference is quite significant as it sheds valuable insight into the types of expression that were used by ancient military forces throughout the Graeco-Roman Mediterranean world: that is, a battle expression could be either traditional or spontaneous in nature and popular in performance.

Military forces of the Graeco-Roman world universally adopted common battle expression types to unite the rank-in-file. Enthusiasm for battle could be measured through mass demonstrations of movement and/or sound. Military forces expressed these common types in their own cultural manner, making them appear unique. Evidence reveals that battle expression types could be traditional, rehearsed undertakings or spontaneous vocal and gesticular displays that were inspired by the battlefield circumstance. Taunting the enemy for morale-boosting satisfaction was a common form of battle expression adopted by all military cultures. The psychological ramifications greatly differed depending on which army was on the receiving end of the taunt. All taunts served to inspire the taunters while attempting to reconfigure the enemy's military preparedness and intent. The manner that an army intentionally presented itself on the battlefield to the enemy should be acknowledged as a common battle expression type integrated within all cultural groups of the Graeco-Roman world. The purpose behind the implementation of the battle expression must be seen from a psychological dimension. Fear and panic could prove contagious and paralysing to an army. The great lengths that armies attempted to instil cohesion and resolve amongst the men to prevent fear and panic from spreading contribute greatly to our understanding of the purpose and nature of battle expression types. The sponsorship of the battle expression by the military high command reveals its military significance in a battlefield environment. The range of battle expression types available to ancient armies and the rehearsal of them beyond the battlefield suggests they were employed as a military tactic that could influence the outcome of the battle. Ultimately, there was an array of battle expression types inherent within every military force of the Graeco-Roman world and were used as an extension of military strategy.

Notes

1 Sabin (2007): 403, 429.
   For the effects of terror at the beginning of battle, see Fron. *Strat*. 2.4.3.
2 Kellett (1990): 215–235.
3 Hanson (1989): 96–151 can be used as a case study for the psychological effects of the Greek phalanx warfare.
4 Kellett (1990): 216.
5 For examples of fear that could develop on the battlefield, see; Hom. *Il*. 13.279–283; Ono. 28.1; Lysias *Mantheos* 16.17; Plut. *Aem*. 19.1–3; Poly. *Strat. Iph*. 3.9.8; Thuc. 5.10.5–7; Polyb. 4.64.9–10, 18.25.1–2.
6 Kellett uses the term "fog of war". This term was first coined by Prussian military analyst Carl von Clausewitz in the 19th century.
7 For evidence of battle expression used in training, see Plut. *Lyc*. 21.1; Ono. *Strat*. 29 "Shouting in the midst of battle"; Amm. 22.4.6; Jos. *BJ*. 3.70–76; Poly. *Strat. Pers*. 4.21.1; Aesch. *Sept*. 270.
8 Kellett (1990): 217, 219.
9 For examples of unit identity, see Hanson (1989): 117–118, 122–124. It suggests the confidence within a phalanx grew out of strong bonds of unit cohesiveness. Hoplites gained courage because of capabilities of the general and the men at their side. Shame of playing the coward and living up to the ideal of the brave man were other determining factors in unit identity within a Greek phalanx army. Ono. 24. Men fought best when brother is in rank beside brother, friend beside friend, lover beside lover; Plut. *Mar*. 19.3–5.

10 Kellett (1990): 226.
11 Lendon (1999): 292.
12 Xen. *Ana.* 3.1.42–44.
13 Lendon (1999): 291.
14 Lendon (1999): 291–292.
15 Xen. *Hell.* 4.8.38, 7.1.31, 7.2.21–22; Xen. *Ana.* 3.1.42 & Lendon (1999): 292.
16 Xen. *Hell.* 4.4.12, 4.8.38, 5.2.42, 5.4.45, 7.5.24; Thuc. 4.96 & Lendon (1999): 292.
17 Hanson (1989).
18 This idea of the individual hoplite having a chaotic experience during battle is echoed by De Vivo (2014): 163–169.
19 Hanson (1989): 100.
20 Hanson (1989): 149 deemed "the ancient equivalent of the rebel yell".
21 See Caesar 3.92 soldiers' triumph songs; sling bullet inscriptions from Olynthus at the time of Philip II; Xen. *Hell.* 7.5.24–27 the impact the death of Epaminondas had on Theban hoplites at Mantinea; Hanson (1989): 107.
22 Krentz (1985): 60.
23 Krentz (1985): 60.
24 Crowley acknowledges the psychological support provided by the phalanx in his monograph *The Psychology of the Athenian Hoplite* (2012).
25 See Poly. *Strat. Iphicrates.* 3.9.8; Xen. Hell. 4.4.12, 4.8.38, 5.2.42, 5.4.45, 7.1.31, 7.2.21–22, 7.5.24; Xen. *Ana.* 3.1.42; Thuc. 4.96; Lendon (1999): 292.
26 Poly. *Strat. Solon.* 1.20.1; Thuc. 5.69–70.
27 Lendon (1999): 296.
28 Caes. *Gal.* 6.39, 7.28; Plut. *Ant.* 39.4; Livy. 6.29; Poly. *Strat. Croe.* 7.8.1 & Lendon (1999): 296.
29 Caes. *Gal.* 6.39; Caes. *B. Civ.* 3.84; Livy. 21.46.6 & Lendon (1999): 296.
30 Lendon (1999): 297 & Livy 21.46, 38.29.
31 Lendon (1999): 298 & Livy. 9.32; Caes. *Gal.* 1.39.
32 Caes. *B. Civ.* 3.92.
33 Caes. *B. Civ.* 3.92; Plut. *Lyc.* 21.1; Ono. *Strat.* 29 "Shouting in the midst of battle"; Amm. 16.12.20, 22.4.6; Jos. *BJ.* 3.70–76; Poly. *Strat. Pers.* 4.21.1; Maur. *Strat.* 2.18, 3.15, 7.15–16, 8.2, 12B.24.
34 For example: Plut. *Lyc.* 21.1.
35 Ono. *Strat.* 29.
36 Veg. *DRM.* 2.14, 2.16, 3.12; Maur. *Strat.* 2.18, 3.15, 7.2.15–16, 8.2, 12B.24.
37 Jos. *BJ.* 3.87.
38 Amm. 22.4.6.
39 Tac. *Ann.* 1.16.
40 Suet. *Nero.* 16.2, 26.2; Tac. *Ann.* 13.25.
41 Plut. *Sul.* 18.3.
42 See the chapter in this book related to "Socio-political and military identity".
43 Suet. *Nero.* 25; Tac. *Ann.* 13.25.
44 Maurice *Stategikon* 2.19. See also Speidel (2004): 110.
45 This notion is supported by Sabin (2007): 406.
46 D. H. *Ant.* 7.4; 9.11; Livy. 10.40, 23.16; 33.9.1–2; Arr. *Ana.* 3.9.7–8; App. *BC.* 2.11.78; Amm. 24.6.11.
    Conversely, there is an element of surprise on behalf of the authors at the lack of battle expression, that would otherwise be customary: App. BC. 2.11.78 & Livy 28.14.
47 Arr. *Ana.* 3.9.7–8.
48 Hdt. 1.17.
49 How & Wells (2008): 74. For a more modern commentary that How and Wells' 1912 study, see Asheri's chapter on Book I in "*A Commentary on Herodotus Books I–IV*" David Asheri, Alan Lloyd, Aldo Corcella; edited by Oswyn Murray and Alfonso

Moreno; with a contribution by Maria Brosius; translated by Barbara Graziosi . . . [et al.]. Oxford: Oxford University Press, 2007.
50 Tac. *Hist.* 1.18; Caes. *B. Hisp.* 25; Thuc. 4.34.1–2.
51 Polyb. 1.32–33.
52 Polyb. 1.45.4.
53 Sal. *Jug.* 60.
54 Sal. *Jug.* 60.
55 Sal. *Jug.* 60.3–4; 98.6–7.
56 Sal. *Jug.* 60.3–4.
57 Sal. *Jug.* 60.3–4. This reference reflects a similar episode in Thucydides' account of the Athenians at Syracuse (7.71) in their failed attempt to capture the city. Here, Thucydides has the Athenian army on land swaying their bodies in torment as well as expressing united cries of success and despair as the army watched its naval forces battle on the sea. The purpose of Sallust recording the African battle expression at Zama is twofold. Firstly, Sallust would have viewed Thucydides version of events at Syracuse as similar in nature to his. By recording similar events to Thucydides Sallust's historiographical ambition to position himself in relation to the well-respected military historian and literary stylist is fulfilled. Secondly, Sallust appears to be recording an extraordinary action committed by a foreign enemy to Rome that helps to not only set the scene but provides insight into Sallust's Roman audience the nature of the enemy in Africa. From a "modern" 21st-century perspective Thucydides' scene aims to generate empathy from his Greek audience towards the Athenian subjects at Syracuse. Thucydides' account of the failed Athenian invasion of Sicily is full of literary creation and as such the historical reliability of this account should be questioned in comparison to the purpose and historical sources available to Sallust in his recording of the Zama engagement.
58 Sal. *Jug.* 60.4.
59 Sal. *Jug.* 60.3–4.
60 Sal. *Jug.* 98.6–7.
61 Sal. *Jug.* 98.6.
62 Plut. *Sul.* 18.3.
63 Plut. *Sul.* 18.3.
64 Ov. *Am.* 3.2.73.
65 Shelton (1988): 354 *n.266*.
66 Plut. *Sul.* 18.3.
67 Caes. *B. Civ.* 3.92.
68 Tac. *Ann.* 14.36; Tac. *Ger.* 7; Caes. *Gal.* 1.51.
69 All ancient Mediterranean cultures held family as their central component. However, not all ancient Mediterranean cultures brought their families or womenfolk to battle, this appears to be a practice employed by Celts and Germans.
70 Caes. *Gal.* 1.51.
71 Livy. 38.17.3–5.
72 Tac. *Ger.* 3.
73 Amm. 19.1.8.
74 Amm. 19.2.6.
75 One could assume for a considerable length of time when you consider: 1. The area of ground between the Persians and the walls of Amida, as Ammianus states that the clashing of weapons took place as the Persians charged at the walls. 2. If this action was spontaneous, it would have taken a few minutes, at least, for the entire Persian force to participate after a minority of the troops would have initiated the act. 3. Spontaneous or planned the Persian troops would have savoured the atmosphere and participation of such an impressive logistical feat that was not negative towards them but would have only served to rouse their spirits for action.

76 Xen. *Cy.* 3.3.58–66.
77 Xen. *Cy.* 3.3.59.
78 For the segregation of home and away supporters in European football stadiums, see *UEFA Safety and Security Regulations 2019 Edition:* Articles 17.01; 27.01.
79 Xen. *Cy.* 3.3.62.

5   Intimidating the enemy

Ancient literary and archaeological evidence reveals that military forces from the Graeco-Roman world intentionally employed a range of battle expression types that aimed to adversely affect the psychological state of the enemy on the battlefield. Characteristics of intimidatory battle expressions were large-scale choreographed demonstrations of massed movement and/or noise as well as taunting that could be inspired from events that transpired on the battlefield. The adoption of intimidatory practices on the battlefield appears consistently in the evidence and universally amongst cultural groups, suggesting it was a typical feature of ancient military life. The purpose of gaining a psychological edge over the enemy by deflating the enemy's mental wellbeing had real military outcomes and should be considered as an extension to military strategy. The sophistication and humour that ancient armies exhibited on the battlefield in their attempts to intimidate or distract the enemy justify the adoption of the battle expression as a replacement for war cry when accounting for the sonic and visual displays of military forces on the battlefield that could transpire before and during military engagement in antiquity.

Evidence reveals that military forces, irrespective of ethnic composition, throughout the Graeco-Roman world, adopted tactics that aimed to intimidate the enemy prior to and/or during battle. What follows are references to battle expression types that employed methods to alter the mindset of the enemy. These examples serve to account for each of the predominant cultures of the Graeco-Roman world that appear within the literary and archaeological records that prescribe to this intimidatory practice. The universal nature of this practice demonstrates that it was a well-established feature of ancient military life.

Xenophon provides an eyewitness account of an Asian battle expression dated from the late 5th to the early 4th century BC. As a member of the Ten Thousand strong Greek military force that subsequently had to flee central Persia, Xenophon was forced to march through the north-western satrapies of the Persian empire to reach safety on the friendly coastline of the Black Sea. In the early stages of the *Anabasis*, Xenophon portrays Persian battle expression as quite loud due to the sheer number of soldiers who were present in the Asian army.[1] Xenophon claims that the Persians performed an uncharacteristic battle expression, in contrast to what he and the Greeks were led to believe would occur from an Asian

DOI: 10.4324/9781003280439-5

force. Prior to the battle of Cunaxa, Cyrus warned the Greeks and the rest of his Asian contingent that the Persian battle expression of his brother's force would be, typically, loud due to the numbers in the royal army and not to be shaken by it. The overly large numbers that made up a Persian royal army exemplified the flamboyant qualities of Asian military custom, namely Persian, on the battlefield. However, Xenophon states (undoubtedly to his own surprise and that of the other Asians in the Cyrus-led army) that the army was deceived by this presumption, and the Persian army opposing them did not perform a loud battle expression as predicted but rather a silent, slow march undertaken in disciplined step:

> For they came on, not with shouting, but in the utmost silence and quietness, with equal step and slowly.[2]

This unexpected presentation negatively affected the psyche of Cyrus' army who were unable to recover from this unforeseen dilemma. The dynamic and deceptive Asian military force of the Persian king adopted, prepared and executed a different battle expression for use against their opposing force. Xenophon suggests the aim of the Persian royal army's adoption of an entirely silent battle expression contrary to common belief was devised to unnerve an enemy expecting something different. In the lead up to a vital military engagement, it appears that slight alterations in tactic, such as the battle expression, could unbalance an opponent and their preconceived notion of the engagement sufficiently that would have surely affected their prepared response – in this instance, the mindset of Cyrus' predominantly Asian force. According to Xenophon, the Greek contingent within Cyrus' Asian army was surprised at the silent battle expression of the Persian royal army, to such an extent that Xenophon deemed it worthy of recording the episode in his history. The degree of close discipline and training required to produce the effect Xenophon describes suggests this battle expression was rehearsed extensively prior to battle. Xenophon's reference demonstrates in this instance that battle expression types were employed to alter the mindset of the enemy and that to achieve success in this dimension of battle required high levels of military planning and preparation from the high command down to the rank and file.

Xenophon refers to the gruesome battle expression performed by a people from northern Asia Minor, the Chalybes. During the march of the Ten Thousand throughout the north-western satrapies of the Persian Empire, the Greeks encountered various Asian peoples and cultural practices. Xenophon recorded the military customs of the peoples that were encountered, such as those of the Chalybes. Chalybean warriors would cut the throats of the enemies killed and then decapitate them. The Chalybes would then carry the heads of their defeated enemies away with them as a trophy. Xenophon claims that the Chalybes would always sing and dance whenever the enemy was likely to see them.[3] The Chalybean battle expression of singing and dancing in front of their enemy appears to have been practiced whenever the enemy could bear witness to it. However, as the reference implies, the Chalybes could not have boasted over their spoils of war (decapitated heads) before the battle began, unless Chalybean warriors kept old trophies with them.

Likely, it appears more probable that the battle expression mentioned by Xeno-phon was reserved for occasions either mid and/or post-battle as the heads of the slain were obtained and a lull in battle may have occurred. In the context of this practice, there may have existed a socio-religious undertone whereby Chalybean warriors defined themselves – their warrior style, military customs and religious obligations to their deities as having to fulfil specific actions, song and dance rituals. Anthropological study reveals that gestures, dancing and song ensure the survival of traditional cultural practice within communities.[4] The Chalybes were demonstratively egotistic in their battle expression, which potentially served to honour their traditional cultural practice at the same time as offending and demor-alising the onlooking enemy with the sights of fallen comrades' severed heads.

A similar battle expression to that of the Chalybes is referred to by Xenophon and sheds greater insight into this practice. Xenophon refers to the Mossynoeci people of northern Asia Minor as being allied to the Ten Thousand Greeks flee-ing Persian territory. The Greek force and Mossynoeci warriors allied together to engage in battle with a hostile Asian force. During the fighting that followed Xenophon records that as the Mossynoeci moved forward to attack the enemy, one of the warriors went forward, presumably a leader of some sort, and then all the rest followed him while they sang a specific tune. After the battle, the Mossynoeci warriors cut the heads off those they had killed and held them aloft to show their allies, the Greeks, and the enemy they had just fought. The display of defeated foes and their heads served to demonstrate the deadly capacity of these warriors to their allies but more importantly their enemies. The intention of this practice aimed to instil trepidation in the mindset of the enemy who bore witness to such sights. Xenophon claims that after showing off the heads of the slain, the Mossyn-oeci warriors all broke into song and danced at the same time to a specific kind of tune.[5] The unfamiliar nature of this battle expression prevented Xenophon from detailing anything more. The similarities between the battle expression of two sep-arate Asian peoples, the Chalybes and Mossynoeci, reveal that this was a practice associated with peoples of the northern Asia Minor region – in this case, flamboy-ant and egotistic in practice, but which may have held a spiritual undertone – the significance of which was known to the participants. The foreign nature of this battle expression inspired Xenophon to record it for his Greek audience.

Xenophon claims that battle expression types were exclusive in nature to the cultural group that performed them.[6] In this instance, the culturally defined battle expression type highlights the significant role intimidation universally had on mil-itary forces. Xenophon refers to a military engagement that saw a Chaldean mili-tary force confront an Armenian force where both sides attempted to seize control over an elevated area of land. The Armenian force moved towards the area of land which was the issue of contention. The Chaldean force, which had arrived on the field before the Armenians, had formed up in battle order and awaited the arrival of the Armenians. Prior to battle, the Chaldeans performed a battle expression "as they were accustomed".[7] This statement reflects the developing sense in the source tradition that battle expression was culturally specific, in purpose and nature, to the military force which performed it. Frustratingly, Xenophon does not provide additional detail about this battle expression, but we could assume that its theme was related to the social, religious and/or political customs of the Chaldeans. It should be borne in mind, however, that the author's intention would appear to be, more than anything else, to provoke ridicule. In recording this feature of the battle, Xenophon creates, by means of juxtaposition, a humorous impression of the military forces involved. Simply put, he refers to the Chaldeans at first raising the customary battle expression and then charging the Armenians who, according to *their* custom, failed to sustain their ranks when the charge came.[8] This reference reveals that through intimidatory practices, military forces could achieve real military outcomes, namely forcing the withdrawal of an enemy force from the field of battle.

The history of the Second Punic War provides insight into the negative psychological effects that battle expression had on military forces who were not familiar with the customs of their enemy resulting in their intimidation. Hannibal's invasion of Italy resulted in Roman armies, often consisting of fresh conscripts and raw recruits, being sent against an enemy that had been hardened by war and confident in nature. According to Livy, Roman forces were psychologically intimidated by Hannibal's Carthaginian forces which highlights the Roman soldiers' lack of disciplined battlefield experience. Carthaginian battle expression types are depicted as being wholly foreign and diverse in comparison to those employed by the Romans. Livy does not delve into detail regarding the nature of Carthaginian battle expression, such as movement and sound, however, the impact it had on the Roman army is telling. In Rome's first military engagement with Hannibal in Italy, near the Ticino River, Livy states that after the Carthaginian force had raised their battle expression, Scipio's spearmen broke ranks and fled.[9] This reference suggests the noise and movement produced by the Carthaginians in the early stages of this battle must have been so psychologically impairing for the Roman army, due to its culturally foreign and tactically unfamiliar nature, that a certain section within the army could not remain in position. The Romans practiced their own battle expression and simulated their own battle scenarios in preparation for battle.[10] As a result, therefore, for sections of a Roman army to flee in the face of an enemy's battle expression must imply that the Carthaginian custom was something the Romans had not expected or encountered previously. Similarly, during the subsequent battle of Lake Trasimene, Livy claims that the Roman army knew they had been surrounded and trapped by the Carthaginian force due to the noise and clarity of the battle expression raised.[11] The Roman soldiers in total confusion and fear turned in different directions to face the enemy that had surrounded them based on which direction the noise of the battle expression was heard.[12] This episode supports the notion that the military high command utilised battle expression to achieve desired military outcomes. At Lake Trasimene, the Carthaginian force sounded their battle expression to inform the Romans of their being surrounded. The noise coupled with the reality of their military predicament confused the Roman army and undermined their previous military strategy.

Livy refers to two other separate occasions in his account of Hannibal's invasion of Italy when Roman military forces broke ranks and fled as a result of the

Carthaginian army's battle expression.[13] In both situations, the Roman army had been outmanoeuvred by Hannibal's military strategy. Livy claims that it was the sound of the Carthaginian battle expression that notified the Romans of – indeed, guided them to – their dire situation. The last episode in Livy[14] that refers to the Carthaginian battle expression suggests one reason why Rome's military experienced such difficulties when they heard the enemy on the battlefield. During the early stages of the Battle of Zama (ca. 202 BC), the Carthaginian battle expression consisted of different languages and peoples originating from different lands. On this occasion, however, the Carthaginian battle expression was weak and not cohesive in comparison to the Roman, suggesting why the Romans ended the day victorious. Livy provides detail of the ethnic composition of the Carthaginian army at Zama. The multicultural nature of Carthaginian military forces was nothing unusual and was the case when Hannibal first invaded Italy. Taken together, Livy's references demonstrate that the Carthaginian battle expression had an intimidating impact on Roman military forces, *except* at Zama when the battle expression lacked cohesion.[15] The reason for this intimidation, then, would appear to reside in the interplay of two related factors: on the one hand, that the Roman forces were not accustomed to such a diverse combination of noise, movement and culture in their preparation or experience in battle; on the other, that, to be effective, this combination of sight, sound and action demanded cohesive performance, derived (one suspects) from diligent training and cumulative field experience. When these factors coincided the culturally foreign and tactically unfamiliar Carthaginian battle expression, which consisted of multi-national language, song, dance and/or noise, appears to have been too disconcerting for some Roman military forces.

Plutarch refers to several battle expression types through his coverage of notable Roman generals' lives and their military campaigns against the Parthians that unnerved the Roman army. These references reveal the intimate understanding ancient cultures, notably the Asian, had with sonic and visual media available to them and their integration into military custom for the purpose of overawing enemy forces opposite them on a battlefield. Plutarch's *Life of Sulla* highlights a Parthian army's intentional use of colour deliberately configured to provide an optical illusion that unnerved men in the Roman ranks on the battlefield. The Parthians are portrayed as ostentatiously displaying their clothing and armour to the enemy in their battle expression.[16] Besides the typical boasting and insults hurled at the Roman army by the Parthians, which Plutarch does not detail at any great length, the main battle expression performed was a display of massed movement. The sheer size in numbers of the Parthian army when combined with the colourful clothing and armour of the soldiers inspired terror within the Roman ranks. According to Plutarch, the Parthians took full advantage of the glittering armour and brightly coloured clothing of their military host by creating an optical illusion that was likened to a flaming fire. The various units within the army surged to and fro in turn which generated a movement of colour that overawed the Roman onlookers. This battle expression appears to have been a rehearsed and pre-planned performance due to the logistics involved regarding the massed movement along the Parthian battlefront and the realisation that the multi-coloured patterns on dress and armour generated the illusion of added movement to an onlooker.

The nature and impact of various Parthian battle expression types are referred to in Plutarch's account of the battle of Carrhae in 53 BC.[17] The first reference highlights the sensory overload the Romans experienced as a result of a brazen and flamboyant Parthian pre-planned battle expression. According to Plutarch, the Parthian military leader at Carrhae, Surena, purposefully concealed the bulk of his army behind his front ranks and he had ordered them to cover themselves with animal skins to hide their armour from the Romans across the battlefield. This act reportedly surprised the Romans as they had expected a more formidable enemy. As the Parthians neared the Romans, Surena gave the signal for the drums, which were intermingled throughout the Parthian line and had bronze bells attached to them, to sound. Plutarch details that the Parthians unlike the Romans used drums rather than trumpets to sound an attack. The sound that the Parthian drums generated was eerie and terrifying to the auditory senses. Plutarch commended the Parthians in his writing for their understanding of tapping into what affects human emotions. Before the Romans could recover from this first sensory assault, the Parthians removed their animal skin cloaks and revealed *en masse* the full array of the army and their glittering armour. The nature of this battle expression which consisted of the playing of drums and removal of cloaks on a given signal suggests high levels of organisation and discipline. Once again, the Parthian understanding of the effect their colourful armour and clothing would have on their enemy was fully exploited in the lead up to this battle and served to demoralise the onlookers. The use of drums and bells, no doubt played to a rhythm, aimed to attack the senses of the enemy, yet the familiarity of the sound to the Parthians would have served to instil confidence within their ranks. As the battle raged, Crassus' son was killed and decapitated and used to instigate another battle expression. According to Plutarch, as the Parthians launched another attack on the Roman lines, they struck up a traditional battle expression[18] and the drums roared again. The ostentatious and brazen displays within Parthian battle expression towards Roman military forces correlate with the overriding characteristics of intimidation tactics aimed at altering the mindset of the enemy.

Celtic and Germanic military forces utilised their physical nature and surroundings to intimidate their enemies. The description of the Germanic/Celtic psyche and physical appearance before and during battle suggests that the character traits of these people were comparable to an untamed beast that is the product of its natural environment. The close affiliation the Germans and Celts had with nature was reflected through the battle expression that they adopted in the lead up to the battle. Moreover, the foreign behaviour of the Celts/Germans was viewed as a curiosity by Greeks and Romans, such as Ammianus and as such highlighted in his historical work.

Livy states that the Celts rank highest in reputation for war due to the purposeful creation of their appearance aimed to generate terror into their opponent before the battle commenced.[19] The Celts' tall physique,[20] their long flowing hair,[21] their large shields and large weapons and their leaping and impersonations of wild animals are listed by Livy as the reason for their military reputation.[22] Tacitus, in his account of Germanic warriors of the Harii tribe, details the appearance of this tribe's military forces and their technique to dominate their opponents. Tacitus states that the Harii in battle blackened their shields, dyed their bodies and selected dark nights and environments to fight their battles in. The result of this battle expression created the appearance of an army that caused their opponents to flee. Tacitus concluded his description by stating that defeat in battle always started with the eyes.[23] Tacitus' reference to the Harii tribe as a ghost-like army has been used, amongst modern scholars, as evidence for the existence of religiously moti-vated bands of Germanic warriors who worshipped Odin/Woden in antiquity.[24] These warrior groups were members of a Woden cult that inspired young men to replicate the *Einherjar*, or dead warriors who entered Valhalla and were chosen to fight in the last battle, Ragnarok. These warriors had an ecstatic relationship with Woden and may have originated to bring reality to Germanic myth.[25] The appearance of the Harii, as a result of this interpretation, suggests they aimed to replicate dead warriors who were still in the land of the living. Alternatively, other modern scholars note that the Tacitus description of a ghostly army was mere lit-erary embellishment.[26] If this description was removed, the Harii would be forest warriors exploiting their native lands for military gain against foreign invaders.

Ammianus Marcellinus, in his experience fighting within the ranks of the Roman military, gives a first-hand account of Germanic battle expressions.[27] These references demonstrate the inspiration that Germanic military forces took out of their animistic culture. Ammianus details the battle of Strasbourg (or Argentora-tum) in which the Western Empire's Caesar, Julian, was attempting to repulse an Alamanni invasion from the Roman province of Gaul. During the opening stages of this battle narrative, Ammianus presents an image to the reader of the German force, gathered before the Roman force, as embracing natural ferocity, as if ani-malistic. Ammianus refers to the Alamannic tribesmen displaying eagerness for battle by grinding their teeth together while clashing their weapons against their shields.[28] This reference is prefaced by the statement that the Germans were alien to the Romans as a result of their wild and barbarous nature that made them prone to readily fall into a mad lust for battle.[29] During the battle narrative, Ammianus compares the German and Roman military forces:

> For in a way the combatants were evenly matched; the Alamanni were stron-ger and taller, our soldiers disciplined by long practice; they were savage and uncontrollable, our men quiet and wary, these relying on their courage, while the Germans presumed upon their huge size.[30]

Ammianus and Tacitus are the literary authorities regarding Germanic battle expression, the *barritus*.[31] The nature of the *barritus* is detailed by Ammianus and Tacitus and as a result, a near-complete understanding of this battle expres-sion has been gained. The only omission from these authors was the actual sound, difficult to reproduce in a literary work. It was a vocal noise made *en masse* that did not use any known word or phrase but was a sound. The *barritus* began very softly and over a long duration increased to the point where it strengthened into a sound that was compared to an explosion, such as a wave crashing against a cliff.[32] In Latin, *barritus* refers to the cry of the elephant,[33] and in Ammianus' work, he details the sound that it made and the impact that it had on those who heard it. According to Ammianus, it was so effective at intimidating the enemy through its imposing sound that it was later adopted by the Roman army and used in battle by troops within the Roman ranks.[34]

The *barritus* was originally employed during a battle where the outcome was yet to be decided. It was used by Germanic forces to swing the battle in their own favour through unnerving the enemy by this mighty sound. Tacitus claims that it was undertaken by German military forces to determine, as if through augury, the outcome of the battle by assessing the conviction of the German troops by the sound they could transmit while undertaking the *barritus*. Tacitus states that the Germans would feel inspired (by the effective noise generated) or alarmed (at the weakness of the noise) through the performance of it.[35] To determine the strength of the noise produced by the fighting force, Tacitus claims that the warriors would raise their shields close to their mouths and create the vocal sound required. The sound would reverberate from the shield and back into the warrior and based upon that assessment the warriors felt inspired or alarmed. The association of the *barritus* to an elephant cry and waves crashing against a cliff, from the Graeco-Roman perspective, propagates the animistic elements within the battle expression of Germanic military forces. Despite the Germanic and Celtic cultures being presented by hostile authors as wild or barbaric, the foresight, rehearsal and meaning of such practices that embraced and attempted to harness the forces of nature that were all mighty exemplifies the intimidatory aspect of battle expression.

Incompetent, comical display

The psychological dimension of the battle expression could take on different aspects aside from intimidation. Literary evidence reveals that ancient armies could utilise battle expression types to alter the mindset of an enemy not through intimidatory practices but through intentional displays of incompetence. Through the display of "inferiority" to an enemy on the battlefield the enemy's psyche would alter to one of a relaxed state resulting in complacency and distraction away from pre-established military orders. Livy provides a reference to a Numidian battle expression.[36] This battle expression does not use words, dancing or hand gestures, as described earlier. Instead, 800 Numidian cavalry contrives, successfully as it transpires, to cause the enemy, a Ligurian army, to lose concentration and be lulled into a false sense of security immediately before engagement as a result of their battle expression. The Numidian cavalry, fighting within the ranks of a Roman army in north-western Italy, were tasked with breaking through a blockade made by a Ligurian military force against the Roman army. The Numidian cavalry purposefully feigned riding up to the Ligurian battle line, as if out of control of their steeds, and then retreated to the Roman ranks. The more the

Numidians "charged", the more out of control and non-threatening they appeared to the Ligurian ranks. The result was that, according to Livy, the Ligurian warriors laid down their arms and sat to watch the ridiculous Numidian cavalry struggling to control their steeds. At the most opportune moment, presumably when many Ligurian warriors had laid down their weapons and broken formation, the Numidians, instead of pretending to regain control of their horses and ride back to the Roman lines, charged through the enemy lines unopposed and rode into the open countryside. Here, they wreaked havoc amongst the nearby civilian settlement, which in turn prompted the Ligurian army to break ranks, thus ending the blockade of the Roman army.

This battle expression is interesting for several reasons. Firstly, the manner by which the large contingent of Numidian cavalry deceived the Ligurian ranks into believing they were not so much threatening, as a source of amusement and contempt is highly suggestive: namely, that this was a Numidian tactic which, given such a large body of cavalry, required a great deal of skill and training in order for it to succeed. Secondly, this battle expression alerts us to the fact that military forces in the ancient Mediterranean world used non-threatening methods, in this case, methods regarded as comical and inspiring ridicule, to distract enemy forces from their military objectives, reflective of sophistication and cunning.

Taunting

The purposeful taunting of enemies on the battlefield was a universal military practice undertaken by each culture in the Graeco-Roman world.[37] The nature of the taunting ranged from spontaneous to pre-planned endeavours that aimed to adversely affect the discipline and mindset of enemy forces. Taunting served to strengthen unit morale within the military force partaking in the heckling, by demeaning an inferior, promoting a sense of superiority and confidence in battle.

Literary evidence

Herodotus refers to the battle expression of a Babylonian military force attempting to stave off the threat of a Persian siege against their city. The Persian king Darius I had led his army to the walls of Babylon and, according to Herodotus, was frequently insulted by the enemy force from on top of the city walls. Here, Herodotus claims that the Babylonians would frequently hurl verbal abuse and direct offensive hand gestures towards the Persian lines.[38] He asserts that these taunts and insults reflected Babylonian arrogance and their belief that the Persians would not be successful in their endeavour to capture the city. Herodotus provides an example of one such insult that one Babylonian was recorded to have cried out to the Persians. "Why loiter there, Persians, and not go away? You will take us when mules give birth".[39] In other words, the Babylonian was questioning both the capacity and the intentionality of the Persian army to pursue its avowed course of action.[40]

Thucydides' narrative of the Athenian siege of Syracuse from ca. 413 BC refers to a Syracusan taunt aimed at the besieging Athenians. According to Thucydides:

> Mounted Syracusan scouts constantly rode up to the Athenian army and amongst other insults asked them: "Are you come to settle yourselves here with us, on land that belongs to other people, instead of resettling the Leontines on their own?[41]

The taunt in this reference relies on a rhetorical question based on the legitimacy of Athens' hostility towards Syracuse. The Syracusan scouts suggest that the purpose of Athens' presence in Sicily was to aid the Leontini, a Greek people,[42] in establishing autonomy for them in their "homeland". Yet, as the scouts insinuate, the Athenians were attacking Syracuse which was not Leontini's land and propose Athenian imperialism was the factor motivating the action.

Taunting on the battlefield was not solely reserved for the enemy, but, as Xenophon details, was also used by a military force to admonish friendly troops for motivational purposes. A Spartan-led army contained units from Mantinea whose infantry had recently been driven off the field by enemy peltasts. Most probably to generate greater spirit and fighting fervour amongst their allies, the Spartans taunted them claiming that they feared peltasts in a similar fashion to children who feared Mormo:[43]

> Lacedaemonians were even so unkind as to make game of their allies, saying that they feared the peltasts just as children fear hobgoblins.[44]

The original Greek word used to characterise the reaction of the Mantinean peltasts, for which the translator interposes "hobgoblins", is μορμόνας (from μορμώ, hideous she-monster, i.e. Mormo). To frighten children into good behaviour, adults used Mormo. Mormo's ugly face and fearful appearance, coupled with an association to eat children, explain her affiliation with goblins.[45] Perhaps the reminder of prior military failures, from their well-respected Spartan allies, served to strengthen the fighting determination of the Mantineans to ensure such disgrace would not befall them in the future.

Arrian refers to a taunting battle expression from a military force that opposed Alexander the Great's army during his conquest of the Persian Empire. A Scythian force, hostile to Alexander's army, is recorded as calling out in their barbarous fashion insulting remarks to Alexander. What is frustrating, however, is that Arrian (albeit because of his sources) does not specify the nature of how the insults were communicated. The fact that the enemy could clearly make out the insult, most probably through translation, via guides or allies, suggests the insults were loud and performed *en masse*. The insult offered up to Alexander was a brazen challenge that he would not dare lay a finger on the Scythian host gathered. The Scythians proudly added that if Alexander were to attack them, he would find them more superior in fighting skill and bravery than to the other Asian forces he had already encountered.[46] No doubt this taunt aimed to boost the confidence of those who performed it.

Another example found in Arrian's *Anabasis* relates to Alexander's siege of the Sogdian Rock. The impressive scale and height of this natural fortress, situated in modern Uzbekistan, prompted the native Sogdians to take refuge on it in their rebellion against the invading Macedonian army. According to Arrian, the Sogdians were confident of the defensive capabilities of this fortress that they taunted Alexander and his men when Alexander's herald offered terms for the Sogdian surrender of the rock. Arrian claims that all the men on the wall facing the herald burst into unified laughter and communicated to Alexander in their native language to find soldiers with wings as no other soldiers could capture the rock.[47] Arrian states that many Sogdians – who took up the cry shortly after Alexander's herald demanded the surrender of the Sogdians inside – undertook the taunt. The Sogdian forces inside the fortress would have realised how difficult it would have been for any attacking force to capture such a difficult geographical location. The Sogdians kept women and children in the fortress, testifying to the confidence the Sogdians had in the rock. The Sogdians most likely did not only take refuge here during the invasion of Alexander but had more than likely taken refuge at this location during times of military conflict for generations. Alexander's forces were probably not the only foreign military force that had attempted to gain control of this fortress. Therefore, the reply given by the Sogdians had more than likely been a reply given to the enemies of the Sogdians throughout their history. The fact that the Sogdians gave this taunt in their native language indicates the cultural significance of this battle expression, as not only did this taunt aim to send a message to the enemy, who would have needed a translation, but would have served to motivate and instil steadfastness within the ranks of the Sogdians who were fighting for their homeland. Ultimately, this Sogdian battle expression can be likened to a traditional response to a demand for the surrender of their mountain stronghold, which would have been familiar to these people.

Livy details the wars that plagued Greece and Turkey during the early 2nd century BC, in his account of Pergamon's war against the forces of Antiochus III ca. 190 BC. Due to the relatively small numbers that had been sent out from the city to fight, Antiochus' forces scorned Pergamon's.[48] The belittlement of numerically inferior forces through vocal taunting, and potentially cat-calling, aimed not only to enthuse the numerically superior force with added confidence of an easy victory but also to degrade the enemy's legitimacy as a credible presence on the battlefield. The volatile nature of this atmosphere would have served to psychologically disable the small force or, conversely, increase its fighting spirit and determination to prove itself in the eyes of those that would underestimate them. In the case of this battle sequence, the carelessness of Antiochus' army and their overconfidence resulted in a stunning victory for the force of Pergamon.

In the case of 3rd-century BC Italian conflicts,[49] prior military engagements were recalled for the purpose of taunting the enemy. Livy claims that military commanders on both sides initiated taunting of the enemy, that no doubt would have filtered through the ranks and used to authorise amongst friendly troops inspiration to ridicule the enemy. According to Livy, the Romans were reminded of the many times the Bruttians and Lucanians had been defeated and subdued by their ancestors. While the Bruttians and Lucanians reminded the Romans of the Roman slaves that had been acquired by Carthage, particularly of their soldiery:

> While the commanders on both sides heaped abuse, the Roman on the Brut-tians and Lucanians, so many times defeated and subdued by their ancestors, the Carthaginian on the Roman slaves and prison-house soldiers.[50]

The impact of this exchange in taunts boosted the resolve of the Roman army that according to Livy:

> Those words at last so fired their courage that, as though they were suddenly different men, they raised a shout again and charged the enemy.[51]

The sting of enemy taunting was undoubtedly made more effective when the forces engaged had previous military and social history with each other. As seen earlier, the relationship between the Greek *poleis* and the Italian peoples, who fre-quently interacted with each other militarily, politically and socially, had intimate knowledge of the shortcomings of their rivals and exploited these on the battle-field. This is comparable to the modern world where football supporters during local derby matches between inter-city or regional rivals; who love to hate each other and know exactly what "buttons" to press, metaphorically, to intimidate and "rattle" psychologically the opposition players and their supporters. The competi-tive and volatile atmosphere generated by derby/rival matches in football stadi-ums would have rivalled the atmosphere of hatred and tension on the battlefield in antiquity between geographical and socio-political rivals.

Roman historian Sallust refers to two separate occasions where a Numidian or composite African (Mauretanian-Gaetulan) military force taunted and insulted a Roman army. Sallust claims that, during the Jugurthine War, a Roman army that was besieging a Jugurthine stronghold was victim to Numidian warriors insulting them.[52] During the day and night, Sallust claims that Numidian warriors patrol-ling the walls would insult the Romans below by calling the Roman commander, Marius, a madman and threatening all the Roman soldiers that they would soon become the slaves of Jugurtha, the Numidian rebel leader. This reference sug-gests that the Numidian warriors understood Latin, the language of the Romans, or at least the Latin required to shout out the taunt. The Roman soldiers outside the walls would have found it difficult to understand what the taunt was if it was not expressed in Latin. This reference suggests that great lengths, particularly by Numidian warriors, were taken to ensure a taunt was understood by an enemy when it was directed towards them. The second reference from Sallust refers to the leader of the Numidian force against the Romans during the Jugurthine War, where Jugurtha taunts a body of Roman troops before battle.[53] Of note, Jugurtha's experience of Roman military action serving under P. Cornelius Scipio Aemilia-nus at the siege of Numantia 134–133 BC[54] should be considered when assessing his use of taunting. Jugurtha's knowledge of the Latin language and Roman mili-tary custom on the battlefield indicates that this battle expression was intended to emulate Roman practice on the battlefield, and no doubt, unnerve the Roman army he was fighting against. The Roman soldiers within earshot of Jugurtha would have been surprised at Jugurtha's familiarity with the Latin language. In this reference, Jugurtha is represented as crying aloud to the Roman soldiers, in Latin, that he had just killed Marius, their leader, with his own hand. According to Sallust, this boosted the morale of the Numidian force, which attacked the Romans with greater fury while the Roman troops entered the fray suffering a significant blow to their collective will because of Jugurtha's news. In fact, the Roman leader Marius had not been killed but was alive. This reveals that ancient peoples were not (as modern interpretations suggest)[55] primitive (tactically or strategically) with respect to the use of their battle expression, but rather should be viewed as adaptive, adept and experienced in utilising a range of expressions as an integral component of their panoply of military tactics.

The physical appearance of military forces, aside from numerical size, was also targeted and used on the battlefield to psychologically destabilise the enemy. According to Caesar, Celtic and Germanic warriors had genetically larger physiques than their Roman counterparts; this subject was customarily used on the battlefield by the Celts to taunt the Romans.[56] In this reference, a Roman army besieged a Gallic stronghold, Caesar records the taunts made by the Gauls stationed on the walls to the Romans outside. The Gauls laughed at the creation of a large siege engine so far from the stronghold's walls.[57] What appears to have amused the Gauls most, that they vocalised to the Roman lines, was not only the distance that such a great engine had been made from the walls but the idea that such small warriors were going to have to relocate the engine up to the walls – the Gauls insinuated the Romans could not do based on their weak appearance in comparison to their own.[58] Whether this taunt was communicated in Latin or the local Gallic dialect remains to be seen. Indeed, it would not have been difficult to employ hand gestures to convey these ideas to a foreign enemy. Whatever the language used in this scenario, the presence of cross-cultural wit and humour in this taunt highlight's clear levels of sophistication present within the ancient battle expression, a concept not embraced with the term war cry.

Plutarch details a taunt made by Teutones warriors towards a Roman army under the command of Marius.[59] Similar to the example given earlier, this Celtic/Germanic force found humour in mocking the Roman army. According to Plutarch, the prospect of invading Roman lands inspired the Teutones to inquire of the Romans, rhetorically, whether they had any messages for their wives and loved ones, as they would soon be with them. This suggestive and provocative question, by way of battle expression, no doubt aimed to draw the Roman army out of their camp into the field for battle. The Roman tactic to construct fortifications and hold up the movements of the invading Celtic/Germanic host frustrated the Teutones, who had previously attempted to attack the Roman camp without success. The belief that the Romans would not dare confront the Teutones in a pitched battle led to the Teutones marching past the Roman camp towards Italy taunting as they went by.

The access to food supplies, especially during siege situations, formed the motivation for taunting the enemy who lacked enough resources. In Caesar's account of the siege of Dyrrachium ca. 48 BC, during the civil war he fought against Pompey, an episode of exchange in taunts took place between soldiers on both sides. Despite Pompey's forces being besieged by Caesar's, it was in fact Caesar's forces that were struggling with resources, particularly grain. The discovery and use of an ersatz ingredient made the making of bread loaves for Caesar's forces possible.[60] During the military stalemate situation of the siege, soldiers on both sides used their turn on frontline duty to taunt the opposition. Pompey's soldiers, who believed the Caesarean forces to be in a predicament of limited grain supplies, taunted the enemy suffering from famine.[61] In response, Caesarean troops threw bread, made using the ersatz ingredients, towards the Pompeian frontline demonstrating the abundance of bread at their disposal, negating the Pompeian claims.[62]

In a similar siege situation during the Jewish War, Josephus described Roman soldiers taunting their starving Jewish enemy during the 70 AD siege of Jerusalem.[63] Roman soldiers, fully aware of the dire food circumstances within the rebel-occupied city of Jerusalem, are claimed to have approached the walls that Jewish rebels patrolled and showcased their abundant supply of food. Whether the Roman soldiers simply held aloft in the air an assortment of food and waved it provocatively towards the Jewish rebels, or whether the Romans simply ate large quantities of food in front of the onlooking enemy is not clear.[64] What is clear, in both passages, were the attempts of the taunters to attack, through satirical means, the misfortune and deprivation of an enemy force. The contemplation and thought process involved, by soldiers within a military force, to target an assumed weakness within the enemy that could potentially alter the enemy's psyche, and the entertainment factor this provided the taunters, was undertaken through the creation of a battle expression.

Types of battle expression that involved taunting could focus on the public belittlement and condemnation of key individuals within an enemy force, such as the commander. The public ridicule of a military leader by an enemy force on the battlefield could potentially result in the army, or the leader himself, wishing to defend the reputation that was under attack. Alternatively, these taunts could serve to drive a wedge between commanders and soldiers if their viewpoints on how to respond to these verbal assaults differed. For example, the army may wish to defend the reputation of their commander by engaging with the enemy in response to ridicule, however, the commander may choose to ignore the heckling and instead pursue an alternative timeframe for battle. This divergence in how to respond to the taunting may result in the army losing confidence in their commander.[65] The repercussions of potential discord between the army and the high command, or off-setting prearranged tactics/strategy, may prove beneficial to the army that initiated the taunt as they were able to successfully force the hand of their enemy into military action they may not have taken had the taunt not occurred.[66] In the course of the siege of Jerusalem, during the Jewish War, Josephus records Titus' attempts to offer peace with the Jews should they surrender the remaining part of Jerusalem that they occupied.[67] The taunts that were given in response by the Jewish rebels to terms of peace and surrender were scathing. According to Josephus, the Jewish rebels verbally abused the Jews within the Roman army, who brought the message before the walls, along with Titus and his father Vespasian. The Jewish rebels claimed that they did not want peace but wanted to incur suffering on the Romans before their ultimate demise:

> To this message the Jews retorted by heaping abuse and retorts of the Jewish leaders. from the ramparts upon Caesar himself, and his father, crying out that they scorned death, that they honourably preferred to slavery, that they would do Romans every injury in their power while they had breath in their bodies.[68]

The response of the Jewish rebels to condemn verbally those in the Roman high command, Titus and Vespasian, along with the Jews who served as the mediators between Rome and the Jewish rebels, supports the above notion. The Jewish rebels wished to force the Romans to act contrary to what they had wanted to transpire. Titus' offer for a peaceful resolution did not comply with what the Jewish rebels wanted, namely death and suffering to all Romans. By responding to the Roman offer of peace, with verbal abuse directed at the men in charge of the Roman army and the men who translated the rebel response to the Roman high command, the Rebels hoped to enact rage upon the Romans, and those that delivered the response, to ensure no peaceful resolution would be contemplated. In this way, the Jewish rebels hoped the Romans would not want peace anymore but war (that is what the Jewish rebels aimed to achieve) thereby affecting the Roman military timeframe.

In Tacitus' *Histories*, details of taunting between opposing Roman armies are recorded. The taunts, similarly, aimed to slander the reputation of the military leaders on the battlefield, while also heckling the composition of the enemy through stereotyping. According to Tacitus:

> Different exhortations were heard. . . . The Vitellians assailed their opponents as lazy and indolent, soldiers corrupted by the circus and the theatre; those within the town attacked the Vitellians as foreigners and barbarians. At the same time, while they thus lauded or blamed Otho and Vitellius, their mutual insults were more productive of enthusiasm than their praise.[69]

The appearance, ethnicity and social status of the soldiers in opposing forces were highlighted and ridiculed, while the commanders were hailed by their respective armies and denounced by their enemy. These taunts aimed to alter the mindset of the enemy, Tacitus claims this created enthusiasm for battle and must have been quite atmospheric with noise, laughter and volatility. Both sides attempted to force the other into undertaking a military action that may have proved disastrous or counterproductive to their overall strategy.

Ammianus details a Persian battle expression that consisted of threatening the enemy. The siege of Bezabde in AD 360 saw the Persians, led by king Sapor II, besiege the Roman-held city. According to Ammianus, during this siege, the Persian forces attacked the city walls and as they did so, loudly threatened the defenders.[70] As mentioned earlier, taunts and threats were a typical feature of battle expression from all cultures around the Graeco-Roman Mediterranean world. However, this reference suggests that the Persian threat to the Romans stationed inside Bezabde was so loud and cruel in nature that Ammianus, who was not present at this siege, recorded it. This means that the second-hand written source or eyewitness account Ammianus used to describe this siege noted it due to its effectiveness. While Ammianus did not record the threat word for word, the comprehension of the Romans inside the walls that the Persians were threatening them verbally suggests that most of the Persian soldiers must have participated in this battle expression. The Persian vigour in threatening the Romans must have been exceptional for it to be noteworthy. As a result, the manner that the Persians taunted their enemy leads to the assumption that this action was pre-planned or rehearsed for the atmosphere it generated to be so memorable in Ammianus' source material.

Verbal taunting and gesticulations were employed by ancient military forces to gain a psychological edge over the enemy.[71] The exchange of abuse and taunting between opposing armies is a strange phenomenon evident in ancient Graeco-Roman literature. Reviling and monomachy[72] were preludes to battles, however, taunting/reviling should *not* be regarded as a feature of primitive warfare involving a lack of military discipline. The suggestion that taunting as a typical military practice should be confined to about the first half of the first millennium BC should be challenged. The combination of reviling the enemy *with* monomachy, or the pursuit of monomachy through reviling the enemy was a typical feature of ancient battlefield practices. It is evident that taunting continued to be a strange military phenomenon recorded in Graeco-Roman literature through to the later Roman imperial age. As explored earlier, the sophistication that ancient military forces exhibited when they taunted their enemies, through wit and humour in specific military situations, does not comply with a lack of discipline and training.

Archaeological evidence

Lead sling bullets, otherwise known as *glandes/glandes plumbae*, reveal archaeological evidence for the use of taunting undertaken by a variety of cultures on the battlefield within the Graeco-Roman world.[73] The archaeological discovery of inscribed slingshot bullets on ancient battlegrounds further supports the tradition of battle expression used during battle. The practice of inscribing of words, phrases and images onto ancient sling bullets is similar to the painted messages on bombs from the modern era.[74] Sling bullet inscriptions can be categorised, similar to battle expression, into different categories based on their epigraphical content.[75] Generally, sling bullets contained inscriptions that highlighted; a personal name:

of the inscriber, the slinger or military commander; the name of a people or a geographical location; a message, normally a taunt, intended for the enemy usually in the form of an exclamation. Each of these categories corresponds with the types of battle expression that could be undertaken by a military force in the Graeco-Roman world. For the purpose of the present discussion about taunting, a focus on those sling bullets that contain messages intended for the enemy, typically by way of exclamations, will follow.

Taunting messages on inscribed lead sling bullets can be considered as a black comedic release of tension and a sense of supremacy from the military force firing these missiles.[76] The feeling of inferiority was, therefore, intended to be felt by the enemy on the receiving end. Examples supporting this notion derive from translated sling bullet inscriptions and some with accompanying images. Sling bullets unearthed from Olynthus, that the Macedonian forces of Philip II besieged in the 4th century BC,[77] contain recurring messages inscribed in Greek that include *nika* (conquer); *papai* (ouch); *dexai, labe* (take it); woe, *haima* (blood) and *trogalion* (a candy or almond or the like). Other examples translated include "an unpleasant gift"; "a sweetmeat"; "eat this"; "hold this too"; "take this" (with a thunderbolt on the reverse)[78] and "seize this". Roman inscriptions[79] involve messages for the people of, or from, a settlement during a siege operation; "a gift for the people of Asculum"; "strike the Picenes" (of Picenum); "runaways, you are doomed"; "an evil to you who are evil"; "our persistence will destroy you utterly" (addressed to the besieged) and "strike Pompeius" (addressed from the besieged to the leader of the besieging force).[80] Sexual metaphors are common themes found in inscribed lead sling bullets.[81] The reference to sling bullets causing childbirth pain and urging the sling bullet to "be lodged well" (with a scorpion image on the reverse) or "impregnate yourself on this" are typical requests found on sling bullets.[82] Sexual superiority is also a theme commonly discovered on sling bullets suggesting triumph; *Tet[o] Octavia [ni] culum* (directive to Octavian's backside) reinforces this idea.[83]

The evidence suggests that taunting was specifically designed to respond to a given situation that may have resulted on the battlefield. Mostly, taunting served to provoke or degrade the enemy to gain a psychological edge over them that may result in a reaction that may not be in the interests of the affected force. For example, taunting aimed to lure an enemy out into a pitched battle when it did not suit them. Taunts instilled anger and frustration in an enemy that may have caused the force to disobey/change orders, such as a besieged/besieging force to lose patience in maintaining their strategy. The provocation of individuals and groups on the battlefield, through public ridicule, aimed to destabilise an army. Taunting the enemy on the battlefield boosted the morale amongst the taunters. Ridiculing a perceived inferior force served to instil greater levels of confidence within the perceived superior force. The calming influence of feeling superior in a military context would have negated any prior apprehension that troops may have had for battle. The imagination and sophistication evident within the different forms of taunting reflects cultural traits and humour not otherwise obtained in a military context. References to taunting are found in all military cultures of the

Graeco-Roman world and reveal that battles could be influenced through demeaning verbal and gestural displays.

Ancient literary and archaeological evidence reveals that the battle expression served as an extension of military strategy to weaken the psychological state of the enemy. The lack of concentration, the spread of fear and uncertainty that could engulf military forces as a direct result of an enemy's sonic and visual battlefield displays was a clear function and use of the battle expression in the ancient world. Spontaneous and choreographed demonstrations that aimed to intimidate and unnerve the enemy were utilised across cultural groups in the Graeco-Roman world. The evidence suggests that on occasion, these practices did contribute to desired military outcomes and victory, such as driving an enemy from a battlefield and breaking through enemy lines. The adoption of taunting enemy forces was a common military practice that functioned on two levels: to destabilise the mindset of the enemy and to boost the feeling of superiority amongst friendly troops. The study of ancient lead sling bullets emphasises the psychological dimension and impact that battle expression had on forces on both sides of a battlefield. The preparation levels that went into effective massed sonic and visual displays and the wit exhibited from military forces that exploited events that transpired on the battlefield in the creation demonstrate the high levels of sophistication that went into battle expression types and the military importance they had on the battlefield.

Notes

1 Xen. *Ana.* 1.7.4; 1.8.11.
   In this case, a Persian army in the sense it was commanded by a Persian, but the army was represented by soldiers from Persia and other friendly Asian nations.
2 Xen. *Ana.* 1.8.11.
3 Xen. *Ana.* 4.7.15–16.
4 Ingold (2002): 698–700.
5 Xen. *Ana.* 5.4.14–17.
6 Xen. *Cy.* 3.2.9.
7 Xen. *Cy.* 3.2.9.
8 Xen. *Cy.* 3.2.9–10.
9 Livy. 21.46.6.
10 Jos. *BJ.* 3.70–76; Amm. 22.4.6.
11 Livy. 22.4.7.
12 Livy. 22.5.1–2.
13 Livy. 25.21.9; 27.1.11.
14 Livy. 30.34.1.
15 Livy. 30.34.1.
16 Plut. *Sul.* 16.2–3.
17 Plut. *Cras.* 26.
18 Plut. *Cras.* 26.
19 Livy. 38.17.
20 Celts/Germans were typically larger in stature than Romans. This is made note of in Caesar *Gallic War* 2.30. "they [Gauls] at first began to mock the Romans from their wall, and to taunt them with the following speeches. 'For what purpose was so vast a machine constructed at so great a distance? With what hands', or 'with what strength did they, especially [as they were] men of such very small stature' (for our shortness of

stature, in comparison to the great size of their bodies, is generally a subject of much contempt to the men of Gaul)".

21 Lobell & Samir (2010): 29.
   Aside from long flowing hair, Celtic males would raise their hair to heighten their stature in society, especially on the battlefield. This is evident in the study of Irish bog body Clonycavan Man by Dr Joann Fletcher, where it was revealed that this Iron Age Celt used a plant resin substance, like a modern "hair gel", imported from France or Spain to erect his hair to enhance his moderate stature.
22 Livy. 38.17.
23 Tac. *Ger*. 43.
24 Simek (2007): 71.
25 Lindow (2001): 104–105; Speidel (2002): 268–269.
26 Rives (1999): 308.
27 Amm. 15.5.22; 31.16.9.
28 Amm. 16.12.13.
29 Amm. 16.12.2.
30 Amm. 16.12.47.
31 Amm. 16.12.43; Tac. Ger. 3.
32 Amm. 16.12.43.
33 Lewis & Short (1879).
34 Amm. 31.7.11.
35 Tac. *Ger*. 3.
36 Livy. 35.11.6–11.
37 Sal. *Jug*. 94; Livy. 37.20; Caes. *Gal*. 2.30, Tac *Hist*. 2.21.
38 Hdt. 3.151.
39 Hdt. 3.151.2.
40 It is a common belief that mules in general have difficulty in producing young success-fully. Herodotus made special mention of this insult as he questioned whether mules could in fact give birth. Later, in Herodotus' work (7.52.7), he claimed that a mule did in fact give birth at Sardis with a set of both male and female genitalia. Herodotus used this as an ominous precursor to Xerxes' planned invasion of Greece.
41 Thuc. 6.63.
42 For a general history of the Leontini in Sicily, see Thuc. 6.3–6.
43 For reference to Mormo or Μορμολύκη, see Strab. 1.2.8.
44 Xen. *Hell*. 4.4.17.
45 Strabo. 1.2.8; Aristoph. *Peace*. 474; Aristoph. *Achar*. 582ff.
46 Arr. *Ana*. 4.4.2.
47 Arr. *Ana*. 4.18.6.
48 Livy. 37.20. Another example of a small numerical force being subjugated to taunting from a numerically superior force, see Poly. Strat. Caes. 8.23.11.
49 Rome fought against the combined forces of Bruttians and Lucanians c.214 BC.
50 Livy. 24.15.7–8.
51 Livy. 24.16.1.
52 Sal. Jug. 94.
53 Sal. *Jug*. 101.
54 App. *Hisp*. 14.89.
55 Such as Whately (2016); Cowan (2007); Rance (2015), who do not acknowledge that spontaneous battle expression was inspired from battlefield scenarios. Modern films including "Gladiator", "Centurion", "Spartacus"; television series such as "Rome" do not present the battle expression, or 'war cry', as sophisticated. See Chapter 1 for more on this.
56 Caes. *Gal*. 2.30.
57 Caes. *Gal*. 2.30.

58 Caes. *Gal.* 2.30.
59 Plut. *Mar.* 18.
60 Caes. *B. Civ.* 3.48.
61 Caes. *B. Civ.* 3.48.
62 Caes. *B. Civ.* 3.48.
63 Jos. *BJ.* 5.521.
64 Jos. *BJ.* 5.521.
65 Or vice versa, the commander may wish to engage with the enemy, but the army may not be prepared to do so.
66 For example, see Fron. *Strat.* 1.11.1. "the consuls on their side feigned a policy of delay, until the soldiers, wrought upon by the taunts of the enemy, demanded battle and swore to return from it victorious".
67 Jos. *BJ.* 5.457–458.
68 Jos. *BJ.* 5.458.
69 Tac. *Hist.* 2.21.
70 Amm. 20.7.5.
71 Glück (1964): 25.
72 Glück uses monomachy, or duel, to mean single combat between representatives of hostile forces on the battlefield. For the purpose of the battle expression study, monomachy, or duel, will be adopted to describe the competitive attempts made by opposing armies on the battlefield to control the atmosphere of the battlefield through undertaking battle expression types. This can be likened to the attempts made by opposing football supporters inside stadiums to drown out their rivals through such actions as: singing, chanting, clapping and movement for atmospheric dominance.
73 Evidence from the Asian, Greek and Roman cultures are widespread, as will be shown. See McDermott (1942): 36; Foss (1975): 30; "It is not necessary, however, to suppose that the use of lead bullets would have been confined to Greeks", Ariño (2005): 233–235; Kelly (2012): 294–295.
74 McDermott (1942): 35.
75 McDermott (1942): 36; Foss (1975): 28; Kelly (2012): 290.
76 Kelly (2012): 290, 295.
77 McDermott (1942): 36–37; Foss (1975): 28.
78 Kelly (2012): 291.
79 McDermott (1942): 36–37.
80 McDermott (1942): 36–37.
81 Kelly (2012): 291–294.
82 Kelly (2012): 293.
83 Kelly (2012): 291–294.

6   The religious dimension of battle in antiquity

Religion inspired battle expression types cross-culturally around the Graeco-Roman world. Individuals and military forces invoked cultural deities and undertook religious customs to prepare themselves for battle. Battle expression types revolved around religious practices such as glorifying deities, invoking the power and wrath of gods associated with war and steeling men in the moments leading up to battle. Ancient literary evidence reveals that religion played a significant role cross-culturally in military life, especially on the battlefield, to the extent that religion and battle expression had an inextricable connection. Through the performance of prayer, hymn, ritual action and/or the imitation of spiritual forces, individuals and armies prepared for battle while expressing their religious belief and identity. The religious dimension of battle expression further contributes to our understanding of the psychological nature of warfare.

Graeco-Roman authors refer to religious-themed battle expression types performed on battlefields by military forces that span from the archaic period to late antiquity. References to individuals and armies worshipping deities that have military association in the lead up to battle is a frequent phenomenon that sheds light on the mental state and temperament of men in their last moments before the violent conflict. While war cries have been characterised in modern-day media forms, such as film,[1] as aggressive incoherent yells and screams the term battle expression presents battlefield customs in a different light. This paradigm accepts that military undertakings prior to battle did not have to be aggressive or intimidating in nature but rather inspiring and spiritual. The singing of religious hymns by hundreds or thousands of men would have been daunting to witness yet inspirational to be a part of. The advent of Christianity in late antiquity did not lead to the end of religion-inspired battle expression. Instead, Christianity was integrated into the ancient military tradition of battle expression, testifying that ancient armies took inspiration from religious invocation.

Homer refers to the Trojan leader, Hector, with whom a specific religious battle expression is linked.[2] The Trojan king's son is referred to as having a great and loud "war cry".[3] There are two occasions where Homer provides added detail about the battle expression of Hector and his forces. The first instance suggests that Hector was strongly supported by both military strength and noise generated by the shouts of his Trojan battalion he led into battle. Homer claims that the

DOI: 10.4324/9781003280439-6

universal shouting noise of the Trojans made Diomedes (Greek warrior-king of Argos) shudder.[4] Homer explains the strength of the Trojan battle expression was due to the Trojans being supported by the Greek warrior deities, Ares and Queen Enyo.[5] Homer states specifically that Enyo was responsible for the noise that was generated in the lead up to the battle. This added detail may reveal the footprint of the Trojan battle expression; namely, invocation of and praise to the war goddess Enyo, or the Trojan equivalent. A later reference to Hector describes him calling upon his fellow Trojans during battle to perform their battle expression. According to Homer, a Trojan force, led by Hector, attacked the Greek encampment on the shoreline where the Greeks had moored their ships. A fierce battle raged between the Trojans and Greeks around the Greek camp. The Trojans managed to fight their way through to the Greek ships on the shoreline, where Homer claims, Hector called upon his fellow warriors to initiate their battle expression. Hector urged the Trojans within earshot to raise their battle expression with one united voice which Hector hoped would spur on the Trojans to continue their onslaught against the Greeks.[6] Hector claimed that Zeus had permitted the Trojans to be in the situation where they could possibly defeat the Greeks who had invaded their lands. The reference to Zeus while Hector incited the Trojans to unite in one voice may suggest that the Trojan battle expression was in some way linked to the supreme god of the Trojan faith. As Homer and/or his audience did not know who this supreme Trojan deity was by name, Homer used the Greek equivalent instead to accommodate for his Greek audience.[7]

Within the Graeco-Roman world, there were different warrior types found in ancient Germanic military forces.[8] Germanic warriors can be categorised into specific warrior groups that were based on totem animals or natural forces that warriors aimed to replicate in themselves on the battlefield. These included warriors who imitated wolves, bears, bucks, naked berserks, ghosts and strong men wielding heavy weapons. These warriors changed their appearance, shape and mentality to instil fear within the enemy.[9] These stylised warriors wore the skins, or likeness, of their warrior style and roused themselves and their fighting group into a fighting frenzy or madness before the battle. Germanic warriors used prescribed and united chanting, dancing, singing and/or mimicking animal noises[10] (depending on what style of warrior they were) to motivate the fighting group into a berserk rage. The lyrical content of the chants and songs performed by Germanic warriors before and during battle was not recorded. However, insight into the possible themes that comprised the various songs, chants and dances of German military forces in battle can be gained by studying the context of these warrior types. Prevalent features found within the characterisation of Germanic warriors consist of the religious worship of the Germanic god Woden/Odin, recounting the deeds of past heroes and the desire to imitate and embody certain animals/natural powers.[11] An example of a themed battle expression that was used by Germanic warriors in antiquity during battle was the *barritus* chant. The *barritus* is likened to the onset of a storm and its subsequent arrival or the surge of a wave and its final crash against a cliff face.[12] Despite the absence of detail relating to movement or lyrics regarding the *barritus*, the theme of generating a natural phenomenon, like the arrival storm or a crashing wave, provides significant understanding of what inspired German battle expressions.

The evidence used to support the categorisation of Germanic warrior styles is focused on literary and iconographic sources. The literary sources consulted date from antiquity to the early Middle Ages. These sources include Roman historians, most commonly Tacitus and Ammianus, along with Dark Age and early Middle Ages Viking and Irish legends and sagas, such as Beowulf. Central to this argument, iconographic evidence forms the basis of depicting ancient Germanic warrior styles. Representations of Germanic warriors survived from antiquity and the early Middle Ages on items such as scabbards, shields, helmets, buckles, rock drawings, gravestones and weapons.

Similarly, iconographical representations of Germanic war dances reveal the religious nature and purpose of these battle expression types. These images include a rock drawing from Sweden depicting a spear dancer, a bracteate medallion from Denmark showing a war-god dancing and a belt buckle from England depicting a weapon dancer.[13] Each dancing warrior from these sources wears a bird-headed dragon-shaped helmet, symbolic of the Germanic war-god Woden and the God is either the warrior presented in each image or the warrior is a devotee to Woden and aims to worship him by dancing in his honour before the battle.[14] Each dancing warrior is portrayed with arms and legs bent as if in a moving/dancing state. Whether these warriors are in the battle line is unclear as they are sole figures in their respective images, unlike the bronze foil depicting the barritus where there are two warriors side by side as if in the battle line. Each figure is clearly holding a weapon and appears to be ready for military combat. These sources do not show German warriors dancing *en masse*; however, these sources do indicate that the god Woden was significant to German warriors in a battle environment. From these sources, it is evident that dancing in a battle scenario was an acceptable form of worship to Woden. The dancing prowess of each warrior aimed to win the favour of Woden and demonstrated their fighting capabilities. The dancing warriors depicted as leaping and jumping in the air while dancing may be compared to the Greco-Roman literary sources that detail the rhythmic leaping and jumping of Germanic warriors before battle.

Germanic warriors undertook a variety of unique battlefield ritual actions that could take the form of song, dance and movement embracing the principles conveyed through the concept of battle expression. Modern scholars frequently use the term "war cry" in their attempts to explain the vocal battlefield customs of ancient Germanic warriors. This includes accounting for singing, chanting and the *barritus* cry.[15] However, these same scholars detail a variety of other battlefield customs that are associated with the Germanic warrior practice such as chanting, dancing, berserk[16] behaviour and the adoption of animal and natural traits for specific warrior styles. As such, using the term "war cry" to fully capture the diversity of ancient battlefield customs available to different cultural groups is limited when confronted by military practices that exhibit diverse elements. The term "battle expression" better serves to embrace all the warrior rituals that may be characterised as similar in military and cultural purpose and meaning. Therefore, if we accept that the practices of German warriors represented cultural beliefs that held socio-religious importance,[17] then it is reasonable to view practices like these as exemplary of what this study of the battle expression in the Graeco-Roman world claims.

In this regard, while beyond the chronological confines of the present discussion, it is interesting to note that Germanic tribes that invaded and settled in the lands of the former Western Roman Empire adopted Christianity as their religion and incorporated Christian sentiment for use in battle expressions. The Norman poet Wace in his *Roman de Rou* wrote a chronicle of the Norman invasion and conquest of England under William the Conqueror. This work was composed in AD 12th century and details the wording of Norman and Saxon battle expression in the opening phase of the battle of Hastings. From Wace's writings, the Saxon and Norman military forces both aimed to conjure the aid of the Christian God in their fight. The Saxons unanimously cried together "God almighty"[18] while the Normans cried together *Dex Aie* or "God help".[19] The Saxons vocalised their intentions for their battle against the invading Normans by crying together "Out!". The public and unanimous invocation of divine powers in battle remained a consistent feature of Celtic and Germanic military forces from antiquity into the medieval period.

Paean

The singing of the *paean*[20] by ancient Greek military forces, before and after the battle, was universally practiced throughout the Greek world.[21] Central to Greek laws regarding warfare related to religious observance. Ancient Greek religion differed from most modern religions in that it was not associated with a creed or fixed belief system. The gods demanded recognition through sacrifice and other ritual acts. The laws of war arising from religious customs involved protecting the property of the gods and ensuring that rites and sacrifices proceeded without interruption.[22] Ancient Greek armies, therefore, undertook battle with a religious orientation, explaining the customary practice of reciting the paean prior to battle. Literary evidence emphasises the pious nature of Greek armies when it came to battlefield custom.[23] Homer refers to the inspiration and resolves Greek troops gained from the gods before the battle.[24] The paean hymns that were offered up to Greek deities, Ares before battle and Apollo after a successful battle,[25] were aimed to invoke the intercession of the deities adhered to.

Evidence reveals that the paean differed, in sound and lyrics, depending on the geographical origin of the army performing it. For example, Thucydides details a battle fought during Athens' failed attempt to capture Syracuse in 415–413 BC between the invading Athenian force and the Syracusans. During this military engagement, the enemy and contingents of allies within the Athenian army sang a similar paean hymn that surprised and unnerved the Athenian army. The surprised and unnerving reaction of the Athenian force reveals that the Athenians were not prepared for this eventuality and might have questioned the loyalty and intention of their allied contingents within their own ranks. The Athenians clearly sang a different paean to other Greek states, and this difference in paean could jeopardise the fighting preparedness of a force that was unaccustomed to it. The unexpected outbreak of an unfamiliar paean, coupled with the loud noise that would have been generated on the battlefield by both the enemy and allied contingents, gives an understanding for Athens' military failure:

> But that which put the Athenians at the greatest disadvantage and did them most harm was the singing of the paean; for the song of both armies was very similar and caused perplexity. Whenever, that is, the Argives or the Corcyraeans or any Dorian contingent of the Athenian army would raise the paean, the Athenians were just as much terrified thereby as when the enemy sang.[26]

Dorian Greeks adopted and implemented a common type of *paean*, which is exemplified in the above extract from Thucydides, whereby the flute and other woodwind instruments were used to accompany the sound of the men singing.[27]

In W.K. Pritchett's study of the Greek military *paean*, he has formulated in sequential order of actions indicating how the Greek marching *paean* on the battlefield may have unfolded:

> The commander-in-chief, whether general or king, gave the command to advance by beginning the paian. The trumpeter sounded the call. The soldiers joined in the song . . . the paian was a sort of hymn or chant. . . . Once the battle was joined, the marching paian might be replaced by the war cry.[28]

It is to be observed that there is a distinction made between the religiously inspired *paean* and the war cry. *Paeans* were commonly followed on the battlefield by wordless chants such as ἐλελεῦ.[29] These chants can be regarded as shouts of joy or confident cries aimed at invoking the gods.[30] That wordless chants contained religious sentiment that can be thematically linked to the religiously inspired *paean* hymn. The *paean* hymn and the wordless chant ἐλελεῦ are two different vocal undertakings. However, they are both used primarily to inspire the men performing them with the belief that certain deities would support their military endeavours. In the same instance, these vocal actions served to create a cohesive fighting force. On another level, both the *paean* and wordless chant served a secondary role: to intimidate the enemy through the creation of loud atmospheric noise and appear before the enemy as a formidable fighting force. This reveals that the war cry concept does not acknowledge the paean hymn as having a role in motivating an army for battle or being intimidatory for the enemy. The misinterpretation of the paean as not being a feature comparable to a war cry highlights the limitations of the term, and the arguments that surround it, to account for sonic and visual battlefield customs.

Xenophon claimed that the singing of the *paean* and raising the war cry were recited prior to battle, elucidating that both are part of a typically Greek battle expression:

At length the opposing lines were not three or four stadia apart, and then the Greeks struck up the paean and began to advance against the enemy. And when, as they proceeded, a part of the phalanx billowed out, those who were thus left behind began to run; at the same moment they all set up the sort of war-cry which they raise to Enyalius, and all alike began running. It is also reported that some of them clashed their shields against their spears, thereby frightening the enemy's horses. And before an arrow reached them, the bar-barians broke and fled.[31]

This extract refers to the Greeks raising the *paean* as ἐπαιάνιζόν τε οἱ Ἕλληνες. The reference to the Greek war cry is οἷον τῷ Ἐνυαλίῳ ἐλελίζουσι. The war cry reference can be reinterpreted through the translation of the original text. Xeno-phon states that the Greeks all cried out ἐφθέγξαντο πάντες.[32] Xenophon used the verb for a chorus singing together, but in this context may be better understood to refer to the production of massed noise, and so does not have to be interpreted as harmonious singing or chanting. Xenophon follows with the raising of a cry "like the kind they shout to Enyalius" (Enyalius being "the Warlike" i.e. Ares). The verb here is an onomatopoeic one (ἐλελίζουσι), suggesting the cry ἐλελεῦ. Based on this summary, it can be deduced that the Greeks shouted loudly, and this was like the sort of wild shouting done when crying out to Enyalius. This is not sing-ing or a hymn or chant as such. It appears that in the period of preparation before the commencement of battle at Cunaxa in 401 BC, where Greek military units fought within the ranks of a non-Greek army. The Hellenes utilised two types of battle expressions – most likely both dedicated to Ares: the *paean* in the first instance, dedicated to Ares;[33] and the second being a massed vocal cry likened to the cry offered up to Ares during worship. This *eleleleu* cry is also mentioned in Aristophanes' *Birds*.[34]

What should be understood from Xenophon's reference is that the separation of *paean* and war cry as being different, which is echoed through Pritchett, should be challenged. Both the *paean* and war cry, in this instance, were similar in religious nature and purpose and should be embraced within the concept of battle expres-sion. The two types of battle expressions noted by Xenophon served to unite and inspire the Greek contingent fighting in the battle along ethnic and nation-alistic lines, by focusing on the supreme war god of their collective faith, Ares. By approaching this extract through the concept of battle expression, the Greek *paean* and war cry are not seen as separate entities but are both categorised as pre, during and/or after battle phenomena that held intrinsic meaning to those that participated in it. War cries used by Greek armies could be inspired by religious sentiment and the *paean* should be considered in the same light.[35]

The singing of religious hymns sought to focus on Greek hoplites for battle.[36] Athenaeus claims that the singing of *paean* hymns exhibited manly vigour, of magnificent bearing that sobered and intensified the individual or group.[37] Thucydides, as seen earlier, refers to opposing *poleis* singing the same *paean* before battle, demonstrating their significance and universal use amongst Greek hoplite armies.[38] The familiarity hoplite soldiers had with singing *paean* hymns, in the theatre and religious ritual outside of military contexts,[39] would have made them an effective form of battle expression by way of sound generated and atmosphere created. According to Pritchett, "the Greeks raised their voices in song at a time when we are told that the enemy would have been taken unprepared if the phalanx had advanced in silence".[40] This is interesting given Pritchett suggests that the *paean* and war cry were different. The *paean* had a tremendous effect on the participants singing and the enemy bearing witness to it, criteria by which, technically at least, a war cry should be categorised.

The singing of the *paean* was not solely reserved for battle scenarios but also held an important function in social gatherings and religious ritual within Greek society.[41] It was customary for Greek military forces in camp before or after mealtimes to offer up sacrifices and libations to the gods followed by the singing of the *paean*. Xenophon details a feast for the Ten Thousand shared with Thracian tribes. The Greek contingent made the customary libations to the gods and followed up with the singing of the *paean*.[42] Xenophon likewise refers to the singing of the *paean* after the pouring out and offering of libations before sentries were posted in the camp and the army went to sleep.[43] Athenaeus details the practice of the Spartan army competing with each other after dinner through the singing of hymns and recital of poetry.[44] *Paeans* were performed in Greek plays and found in poetry too.[45] *Paeans* were frequently rehearsed and practiced outside of battlefield scenarios. Literary evidence reveals that Greek hoplites were well acquainted with the practice of singing *paean* hymns. The effect that the singing of the *paean* had on the battlefield would have created an inspiring – for the participants – yet solemn – for the enemy – collocation through sight and sound of military unison and commitment. The Greek military custom of singing the *paean* before the battle had an adverse effect on non-Greek military forces, which suggests these non-Greeks were unaccustomed to this type of battle expression. Xenophon refers to Asian military forces taking flight in repeated battles after the Greek forces opposing them had sung the *paean*.[46] The evidence suggests that the *paean* sung by a Greek army in battle was a cohesive and effective performance that derived from the familiarity hoplites had with it, unique to the Greek *poleis*.

Paeans were sung to avert danger or disaster, often in a sacred context, to accompany sacrifice.[47] It is known that Greek armies universally offered sacrifice before battle as the *paean* was sung. In the case of Spartan armies, a she-goat was customarily sacrificed before the battle.[48] As opposed to *paeans* recited in a non-military context, where singing and dancing would take place, in the absence of dancing in the period of preparation prior to battle, a soloist or leader would sing the *paean* song, while the refrain, or chorus, was sung by the rest of the rank and file.[49] Communal singing was an important part of the fabric of Greek military life, particularly the recitation of Tyrtaeus' poetry in Sparta.[50] The *paean* should be regarded as a significant battlefield custom that correlates with the paradigm of the battle expression. The adoption of a socially familiar custom, such as singing the *paean*, and implemented, with alterations, for effective military purposes supports this argument.

The *paean* was a short prayer to Apollo which commemorated Apollo's fight with the Delphic dragon when Apollo was encouraged to shoot it using a bow and arrow.[51] Apollo's connection with men, particularly young men, through education of the arts, physical training for the military and their initiation into adulthood[52] denotes the reason why this deity was used by hoplites throughout the Greek world as a means for inspiration and intercession before the battle. The *paean* cry to Apollo was generally reserved for the aftermath of a battle that resulted in victory; however, when its origins are placed in their original context, Apollo's *paean* may just have as much relevance for a pre-battle custom, due to his association with battling dragons.

It appears the importance of singing the *paean* in the lead up to battle and in victorious aftermath for Greek armies was profound. *Paeans* were used for two main reasons: to seek the intercession of the gods by honouring them and recalling notable exploits and deeds, such as Apollo's slaying of the Delphic dragon; and in conjunction with the offering of a blood sacrifice before the battle. Ritual action would have served to kindle resolve and courage in the hoplites preparing for battle. Likewise, the unity and effectiveness created through singing together as a military force by way of atmosphere generated, whether pious/reverent or resounding in noise, would have filled the hoplites participating with motivation. Singing these hymns as a military force were useful training exercises for hoplite warfare and this rehearsal contributed to its effectiveness on the battlefield. The use of the *paean* by Greek military forces in the lead up to battle aimed to remind hoplites of the music, dance and movements exposed to in training. This would have helped to create a general sense of order in the battle line and focus the mindset of the hoplites on the battle at hand.[53] Finally, in relation to the *paean* reserved for Apollo, during the classical period the cult of Apollo, particularly, played a special role in the life of the *polis* all throughout the Greek world.[54] Within the *polis,* there were groups dedicated to Apollo and/or the recitation of Apollo-inspired *paeans*, often associated with male initiation. Therefore, the relationship citizens, within the military, had with Apollo outside of the military context reminded them of the *polis* community and their socio-political orientation. The reminiscence of home and the *polis* community further served to galvanise the men within the rank and file to fight for the protection of their land and people before and after the battle. The military dimension of the Greek *paean* correlates with the principles of the battle expression and further highlights the need to replace the war cry as a term to represent this ancient military phenomenon.

The lyrics of the Homeric *Hymn to Pythian Apollo* reinforce the ritual obligations and military traits that Apollo encapsulated, which would have been an attractive source of inspiration for a Greek army on a battlefield.[55] Apollo is presented as a skilled hunter and restorer of justice and peace:

> Nearby is the fair-flowing spring where the lord, the son of Zeus, shot the serpent from his mighty bow, a great bloated creature, a fierce prodigy that caused much harm to people in the land – much to them, and much to their long-shanked flocks, for she was a bloody affliction.[56]

The hymn details the ritual actions required to invoke the benevolence of Apollo, specifically the singing of the *paean*.[57]

The Homeric *Hymn to Ares* gives insight into the sentiment that *paean* hymns, dedicated to Ares, may have contained and why they were used on the battlefield. The lyrics of this hymn clearly attest to the portrayal of Ares as a supreme war deity:

> Ares haughty in spirit, heavy on chariot, golden-helmed; grim-hearted, shield-bearer, city-saviour, bronze-armoured; tough of arm, untiring, spear-strong, bulwark of Olympus; father of Victory in the good fight, ally of Law; oppressor of the rebellious, leader of the righteous; sceptred king of manliness.[58]

The affiliation Ares has with confidence, manliness, strength, resolve and victory for the just in the Homeric hymn exemplifies the connection Ares has with battle-field endeavours. Unexpectedly, in the same hymn, Ares is presented as a deity that is associated with peace and the avoidance of violence.[59] Perhaps the paean hymn to Ares similarly served to invoke the deity to prevent violent confrontation on the battlefield, attributes not widely associated with him in popular culture.

Rome

The notion that Roman armies customarily utilised religious inspired battle expression is evident in Plutarch's *Life of Numa*,[60] which details the foundation and customs of the *Salii* priests of the early republic.[61] Plutarch records the lyrics of a hymn that the *Salii* are claimed to have sung as they performed their annual war dance through the streets of Rome. A subject within the hymn is the craftsman Veturius Mamurius, who helped forge the ancile. Plutarch challenges the lyrics of the hymn by suggesting that the reference to Veturius Mamurius may be inaccurate. Rather the lyrics may have instead mentioned *veterem memoriam*, which is claimed to mean ancient remembrance.[62] The notion of maintaining ancient customs in a military context, evident in the lyrics of the Salian hymn, is consistent with Caesar's claims of war cries being an ancient military phenomenon. Our understanding of the Roman battle expression, and its various forms, must recognise that traditional customs were visible and commonplace on a Roman battlefield.

Forms of battle expression used during the republican[63] period through to the late empire reflect a range of Roman traditions. Practices designed to unite soldiers on the battlefield and intimidate the enemy opposing them appear to have had origins from Rome's early foundations. Evidence for the precise origins of those practices, which Caesar refers to as ancient institutions, is difficult to find prior to the 1st century BC.[64] Archaic practices that are relevant to the development of particular forms of battle expression that contain a uniquely Roman cultural tradition include divine invocations (e.g. to Bellona); the Salian hymn which encompasses Salian customs; chants and songs and Christian traditions.

Bellona

The Roman war goddess, Bellona, was closely connected to Roman military religious life and battle expression.[65] Inscriptions containing reference to Bellona within military contexts, particularly in conjunction with Mars and virtus, have been unearthed in Britain, North Africa, France and Germany.[66] Varro claims that the Latin word for war, *bellum*, has close ties with the Roman war goddess Bellona,[67] attesting to her military significance within the Roman culture. The origins of Bellona worship are unclear and appear to reside in the early republic. What is known is that Bellona assimilated with Cybele and Magna Mater in early imperial times.[68] Livy records that in ca.340 BC, a state pontiff within the ranks of the Roman army called upon a host of Roman war gods publicly, including Bellona, to support their endeavours which resulted in the rise in the morale of the army influencing their victory.[69] Of note, this reference refers to M. Valerius, a public priest, being purposefully called upon by the consul Decius to invoke the Roman gods. The presence of a public priest in the forward ranks of the army for the purpose of calling upon the cultural deities suggests a long-standing connection between the Roman army, state religion and battle expression:

> In the confusion of this movement Decius the consul called out to Marcus Valerius in a loud voice: "We have need of Heaven's help, Marcus Valerius. Come therefore, state pontiff of the Roman People, dictate the words, that I may devote myself to save the legions." The pontiff bade him don the purple-bordered toga, and with veiled head and one hand thrust out from the toga and touching his chin, stand upon a spear that was laid under his feet, and say as follows: "Janus, Jupiter, Father Mars, Quirinus, Bellona, Lares, divine Novensiles, divine Indigites, ye gods in whose power are both we and our enemies, and you, divine Manes, – I invoke and worship you, I beseech and crave your favour, that you prosper the might and the victory of the Roman People of the Quirites, and visit the foes of the Roman People of the Quirites with fear, shuddering, and death. As I have pronounced the words, even so in behalf of the republic of the Roman People of the Quirites, and of the army, the legions, the auxiliaries of the Roman People of the Quirites, do I devote the legions and auxiliaries of the enemy, together with myself, to the divine Manes and to Earth." . . . At the same time the Romans – their spirits relieved of religious fears – pressed on as though the signal had just then for the first time been given, and delivered a fresh attack.[70]

The dedication of the temple of Bellona in Rome in ca. 296 BC can be categorised as a battle expression by way of an oath made to the goddess on the battlefield. Livy claims that in a battle against the Etruscans and Samnites Appius, the Roman commander raised his hands in prayer beyond the standards in front of the army. He publicly vowed to construct a temple to Bellona should the Romans be victorious in the upcoming battle.[71] Through the course of the battle, the Romans were highly successful in driving their enemies from the field. As Appius led his men forward, Livy claims that he called from time to time on Bellona, goddess of victory. With the cry of Bellona, goddess of victory the soldiers' enthusiasm grew.[72] The use of Bellona as a subject for battle expression reveals the confidence Roman soldiers gained from their belief in her benefaction.

Mars and the Genii

The Roman army sought the intervention of certain deities that were specifically associated with military features and were culturally Roman.[73] What appears to be consistent in the Graeco-Roman literary record is the importance of the war god Mars and the *Genii* in military culture. Roman military forces, pre-Christian dominance of the late empire, were closely attached to Mars and the Genii and served as great inspiration for their battle expression. The Roman connection to the gods is evident in the institutions established by Romulus who:

> Recognized that good laws and the emulation of worthy pursuits render a State pious, temperate, devoted to justice, and brave in war. He [Romulus] took great care, therefore, to encourage these, beginning with the worship of the gods and genii. He established temples, sacred precincts and altars, arranged for the setting up of statues, determined the representations and symbols of the gods, and declared their powers, the beneficent gifts which they have made to mankind, the particular festivals that should be celebrated in honour of each god or genius, the sacrifices with which they delight to be honoured by men, as well as the holidays, festal assemblies, days of rest, and everything alike of that nature.[74]

Romans believed that Romulus and Remus were the sons of Mars, therefore, establishing a foundational connection between Mars and the Roman culture.[75] Mars was associated with Roman military life and presided over battles:

> The Sabines and the Romans . . . give to Enyalius the name of Quirinus, without being able to affirm for certain whether he is Mars or some other god who enjoys the same honours as Mars. For some think that both these names are used of one and the same god who presides over martial combats; others, that the names are applied to two different gods of war.[76]

That Mars had a close connection with Roman military practice is clear. Augustus received captured standards from the Parthians, which had been won in battle decades previously. The diplomatic success of this event was comparable to a military victory. Indeed, in honour of this success, Augustus commanded that sacrifices be decreed and, likewise, a temple to Mars Ultor be dedicated on the Capitol, in imitation to that of Jupiter Feretrius, in which to offer the standards; and he himself carried out both decrees.[77] Likewise, the dedication to Mars of a Roman commander's sceptre and crown, worn after a military triumphal procession through the streets of Rome, reveals the association Mars had with the

Roman military; in this case post-battle.[78] The military association with Mars continued into AD 4th century as Mars worship within a battle atmosphere was practiced.[79]

The religious dimension of Roman battle expression can further be understood through the cult of the *Genius* in the army.[80] The origins of the cult of the *Genius* are archaic and there is little doubt that the *Genius* was one of the oldest features of Roman religion; its derivation from the words *gignere* and *gens* attests to this. The army was at the fore of the *Genius* of the Emperors cult – no other manifestation of Roman life left more remains of the cult of the *Genii* than the army.[81] *Genii* worship was embedded within the army, all units in the army had their own *Genii*.[82] Archaeologically, the largest number of chapels, altars and statues unearthed in a Roman military context is dedicated to the *Genius centuriae* in the legions and praetorian guard.[83] This is the case due to the soldiers' strong attachment to their *centuriae* which instilled them with a feeling of identity and belonging.[84] The discovery of religious dedications to the *Genii* inside and outside military camps suggests that the worship of *Genii* may have been a typical feature of Roman battle expression on the field of battle. It could not be remiss to envisage in the lead up to battle different military units, such as the *centuriae*, offering up prayer and dedication to their *Genii*, or from the legion as a whole. This is reminiscent of an episode from Caesar's *The African War*, where a soldier proudly professed his origins as a veteran of the tenth legion; *sed de legione X. veteranus*.[85] Similarly, the words of encouragement prior to battle from Cerialis, during AD 1st-century Roman civil wars, aimed at provoking the pride and spirit of the legions under his command:

> He applied the proper spur to each of the legions, calling the Fourteenth the "Conquerors of Britain," reminding the Sixth that it was by their influence that Galba had been made emperor, and telling the Second that in the battle that day they would dedicate their new standards, and their new eagle. Then he rode toward the German army, and stretching out his hands begged these troops to recover their own river-bank and their camp at the expense of the enemy's blood. An enthusiastic shout arose from all.[86]

That the *Genii* were intrinsically associated with battle expression is clear; the Roman war gods had their *Genii* too, and even the military standards had their *Genii* oath of service worshipped as a deity by soldiers.[87] Roman battle expression embraced the worshipping of origin (name, number, symbol, honour, decoration) of a legion or unit within it. The evidence suggests that the *Genii* were key motivators of Roman armies on the battlefield. This sentiment is echoed by Ammianus, who in the late empire, referred to the *Genii* as being present on the battlefield with men as they fought and were perceived to have been the forces that protected due to the link between the *Genii* and men's souls:

> it was not the gods of heaven that spoke with brave men, and stood by them or aided them as they fought, but that guardian spirits attended them . . . these

spirits are linked with men's souls, and taking them to their bosoms, as it were, protect them.[88]

The significance of Mars and the *Genii* to the military from the republic to the late empire is clear. The belief that the outcome of battles could be decided upon by the intervention of these spiritual forces is reflected in the forms battle expression could take. Literary and archaeological sources suggest that the worship and invocation of Mars and the *Genii* was a long-established military practice.

The Salii

The Salian priests of Rome should also be considered when dealing with Roman battle expression. The origins of this order of priests are believed to have occurred during the reign of King Numa ca. 715–673 BC. The *Salii* were renowned for their dancing and singing of hymns in praise of the gods of war. Their role, besides, was to house and care for the holy relics or ancilia (small shields) in which one fell from heaven as a gift to the Romans for protection.[89] The craftsman Veturius Mamurius fashioned multiple other shields that were identical in appearance to the one that fell from heaven, as Numa was keen to avoid the original being stolen. The priests every March would gather armed with spear, dagger and shield and journey through the streets of Rome dancing, by way of rhythmic leaping:

> For they execute their movements in arms, keeping time to a flute, sometimes all together, sometimes by turns, and while dancing sing certain traditional hymns. But this dance and exercise performed by armed men and the noise they make by striking their bucklers with their daggers, if we may base any conjectures on the ancient accounts, was originated by the Curetes. I need not mention the legend which is related concerning them, since almost everybody is acquainted with it . . . This dancing after the manner of the Curetes was a native institution among the Romans and was held in great honour by them.[90]

The connection of the dancing *Salii* to the *Curetes* is steeped in Graeco-Roman mythology.[91] Of note, in the extract given earlier, is the reference to traditional hymns and the noise made when the Salii struck their shields with their weapons. The lyrics of the Salian hymn have partially survived antiquity through the work of Marcus Terentius Varro in his work *The Latin Language*. Despite being incomplete, the surviving lyrics reveal a close affinity to culturally Roman deities, namely Janus:[92]

> In the Hymn of the Salians: O Planter God, arise. Everything indeed have I committed unto (thee as) the Opener. Now art thou the Doorkeeper, thou art the Good Creator, the Good God of Beginnings. Thou'lt come especially, thou the superior of these kingship . . . Sing ye to the Father of the Gods, entreat the God of Gods.[93]

The archaic origins of the *Salii* and their customs reinforce the notion that forms of battle expression were culturally significant and were inspired by uniquely Roman religious dimensions. Janus being the main subject in the lyrics to the Hymn of the Salians supports this. Whether this hymn was sung on the battle-field is doubtful given the lack of evidence to prove it, however, the Roman "war cry" as referred to by modern scholars and translators of Graeco-Roman literary works, may very well have incorporated elements of this religious tradition. What is certain is that a typical battle expression used on the battlefield prior to engage-ment with the enemy was the clashing of weapons against shields. This practice can be directly linked to the *Salii* whereby they danced through the streets of Rome clashing their weapons against their ancilia. As will be presented, the clash-ing of weapons against shields was a means to invoke the gods, potentially Janus individually and/or the Roman war gods – including, but not solely, Mars.

In his narrative of the battle of Zama, Polybius states that a typical military practice prior to battle was reminiscent of Salian custom. The battle expression involved soldiers creating massed vocal noise in unison while clashing their swords against their shields.[94] What is significant in this reference is the Ῥωμαῖοι κατὰ τὰ πάτρια which alludes to this practice being traditional of the Roman army. Unfortunately, the detail regarding the lyrics or vocal noise that was made by the Romans is not provided by Polybius. However, the clear description of weapons being struck against shields can be linked directly to the religiously inspired and military contextualised custom of the *Salii* who honoured the gods of war by the same action.

Massed vocal noise coupled with the clashing of weapons against shields is recorded as being typical during the early republican period, centuries before the battle of Zama.[95] In a battle narrative described by Dionysius of Halicarnassus, the author claims that the Roman army anticipated a night attack on their entrench-ments by their enemy, the Hernici. Imagining noise made in the darkness was the Hernican army, the Romans:

> took up their arms once more, and forming a circle about their entrenchments, for fear some attack might be made upon them in the night, they would now make a din by all clashing their weapons together at the same time and now raise their war cry repeatedly as if they were going into battle.[96]

Another example of this type of battle expression derives from the early repub-lic where the Roman army fought against the Volsci. The Volscian armies were claimed to have been thrown into confusion at the first onset of the Romans and were unable to endure either the massed vocal noise or the clash of their arms.[97] Plutarch, in his *Life of Antony*, refers to a military engagement against the Parthi-ans. In the opening sequence of this battle, the Romans cried out and clashed their weapons together.[98] The result of this forced the Parthians to flee the battlefield before coming to grips with the Romans.

Roman military forces traditionally clashed their weapons against their shields (pilum against scuta).[99] This was a typical feature of Roman military action prior to battle during the republican period. Roman war cries embraced foreign customs, such as the *barritus*, as the Roman empire expanded and absorbed different cultures within the military. However, the traditional war cry of clashing weapons against shield remained into the early and later imperial age.[100] The late empire's incorporation of Christian invocations into pre-battle custom did not result in the negligence of the archaic religious practice of clashing weapons against shields. Perhaps the military advantages gained because of this tradition superseded the potential heresy that may have been associated with this ancient cultural practice.

Roman armies went into battle noisily or silently, based on the circumstances of the battle.[101] This understanding does correlate with the battle expression paradigm that suggests there was a range of battlefield customs available to a military force within the Graeco-Roman world. Ross Cowan and Adrian Goldsworthy refer to war cries being undertaken by Roman armies and individual soldiers in battle. Both debate whether Roman armies of the mid-republic advanced into battle clashing weapons against shields or if this practice was replaced with a silent advance into battle, before engagement with the enemy and then undertaking a war cry, of the late republic.[102] There is little acknowledgement of what the clashing of weapons against shields may have been significant for aside from the purpose of frightening the enemy and inspiring fellow troops. The purpose of silent advances lies in discipline and strategy.[103] Both viewpoints do not account for the probable religious significance that the clashing of weapons against shields had in a comprehensive manner, hence the need for a revision of the term war cry.

The clashing of weapons against shield simultaneously with some type of massed vocal chant/cry/shout (of which the lyrics and tune are not known) was a culturally Roman battle expression. This military practice had direct links to the establishment and customs of the Roman *Salii*. Different authors writing about separate military engagements across alternate periods of time describe similar battle expression. In each reference, details of lyrics, or noise, generated by the Romans as they clashed their weapons are not provided. However, the act of clashing weapons against shields reveals a clear institutionalised military practice that had been used by different Roman military forces spanning centuries. The cultural ownership of this form of battle expression, which was different from their non-Roman enemies, demonstrates the limitations of the modern understanding of the term war cry and the need for its re-conceptualisation through the notion of battle expression.

During the late empire, the rise in Christian influence in Roman society assimilated into military custom.[104] Maurice's *Strategikon* refers to the Christian battle expression *Deus nobiscum*[105] (God is/be with us) and *Adiuta, Deus*[106] (God, help us) that were in official use during the late empire. The shift in battle expression is exemplified when dealing with the cavalry units,[107] Maurice recommended that the battle cry, *Nobiscum*, should be avoided due to the disruption and unevenness it may cause the cavalry battle line. Instead, Maurice urged the completion of prayer in camp on the day of the battle, and for all in the army, led by the general, priests and officers, to recite numerous times the *kyrie eleison* (Lord have mercy) in unison. Silence was recommended on the battlefield, for clear and effective issuing of orders and maintenance of formation. Only as the frontline closed with the enemy, it was suggested, for the rear ranks to let out a cheer or shout, to unnerve the enemy and bolster the confidence of the troops. During the battle, Maurice suggests that the second line, not committed to battle, should let out two or three rousing cheers to encourage their fellow troops and intimidate the enemy.[108] Despite the spread of Christian ideology and practice in the Roman army, it appears that elements of the *barritus* cry was still in use. The mid-battle rising noise of the *barritus* cry is similar in nature and purpose to the instruction given to the troops in the second battle line, evidence that the tradition of maintaining a bygone battle expression was still employed in the AD 6th century.

Graeco-Roman military forces sought religious inspiration prior to the onset of combat and battle expression types reflected this. Military units and entire military forces adopted religious traditions such as hymn-related and ritual action to invoke divine benefaction and spiritual resolve before the battle. The fusion of religion and military practice is evident within the Germanic, Greek and Roman battle expression types. These cultures consistently oriented battlefield customs along religious lines that served military functions such as venerating deities associated with protection, military prowess and raw power.

Notes

1 Modern scholarly works such as Whately (2016); Cowan (2007); Rance (2015) who do not acknowledge the features that encompass the battle expression definition. Modern films including "Gladiator", "Centurion", "Spartacus"; television series such as "Rome" do not present the battle expression, or "war cry", as diverse. See Chapter 1.
2 It should be noted here that, at best, Homer's depiction of warfare may tell us something about contemporary expectations of battle. On this occasion, the influence religion had on battle expression. Indeed, Homer as a source for Hector urging on the Trojans as evidence for a Trojan battle expression is ludicrous.
3 Hom. *Il.* 5.590–592; 15.671; 15.716–720.
4 Hom. *Il.* 5.595–597.
5 Hom. *Il.* 5.590–595.
6 Hom. *Il.* 15.716–720.
7 Homer is our only source for Trojan beliefs, and it is, of course, possible that the Greeks and Trojans shared a common pantheon of gods. With that in mind, it is also important to remember the fictive nature of Homer's epic and this perspective can only remain speculative.
8 Speidel (2004).
9 Speidel (2004): 45, 69, 82, 110 and 111. See also, note 40 above.
10 Speidel (2004): 45, 69 and Chapter 10 entitled "Chanting".
11 Speidel (2004): 14, 15, 31, 43, 44, 45, 69, 70, 73, 94, 110–112.
12 Speidel (2004): 110–112.
13 Speidel (2004): 118, 119 and 121. Figs. 11.1, 11.2 and 11.3.
14 Speidel (2004): 112.
15 Speidel (2004): 110–113 Chapter 10 on "Chanting".
16 Beserk refers to the manipulation of mind and body to prepare it for violent confrontation that created an adrenalin fuelled state of rage and fearlessness that has been associated with Germanic and Viking warriors.
17 Speidel (2004): 126.

18 Wace. Rom. 19.
19 Wace. Rom. 20.
20 For a discussion on the terms, and their interpretations, "Paean" as opposed to "Paian", see Ford (2006): 277–295.
21 Pritchett (1971): 105; Rutherford (1994): 113–116; Haldane (1965): 33 n.5; 76; Thuc. 1.50, 2.91, 4.43, 4.96, 7.44; Arr. *Ana.* 1.15.7–8; Xen. *Ana.* 1.8.16–19, 4.3.18–19, 4.3.29, 4.3.31, 4.8.16, 5.2.13–14; Xen. *Hel.* 2.4.17, 4.2.19; Xen. *Cy.* 3.3.58; Aesch. *Pers.* 384–395.
22 Lanni (2008): 476.
23 Thuc. 6.69; Hom. *Il.* 11.10–16; Arr. *Ana* 1.15.7–8; Poly. *Strat. Alex.* 4.3.5ff; Curt. 8.11.22–25; Xen. *Ana.* 4.3.18–19; Pritchett (1971): 109.
24 Hom. *Il.* 11.10–16.
25 Poly. *Strat. Solon* 1.20; *Iphicrates* 2.9.7; Xen. *Ana.* 1.8.16–19, 5.2.13–14; Xen. *Hell.* 2.4.17, 4.3.21; Arr. *Ana.* 1.14.7.
26 Thuc. 7.44.6.
27 Pritchett (1971): 107.
28 Pritchett (1971): 107. Pritchett uses Xen. *Ana.* 1.8.18 to support this idea.
29 Haldane (1965): 33, 35 n.17.
30 Aesch. *Pers.* 384–395.
31 Xen. *Ana.* 1.8.17–18. Xenophon's description in the Anabasis of the Greek mercenaries scaring/impressing the "barbarians" with their professional display of ritual drill (rather like a karate expert impressing novices with a display of kata) is of interest and reveals the familiarity the troops had with this battle expression.
32 Xen. *Ana.* 1.8.18.
33 Poly. *Strat. Solon* 1.20; *Iphicrates* 2.9.7; Xen. *Ana.* 1.8.16–19, 5.2.13–14; Xen. *Hel.* 2.4.17; Arr. *Ana.* 1.14.7.
34 Aristoph. *Av.* 364.
35 For more on religious ritual in Greek warfare, see *The Oxford Handbook of Greek and Roman Warfare.*
36 Rutherford (1994): 113–116.
37 Ath. *Dei.* 14.624.
38 Thuc. 7.44ff.
39 OCD 3rd ed. (1997): 1060 "*Paean*"; Rutherford (1994): 113–115.
40 Pritchett (1971): 105.
41 Hornblower, Spawforth & Eidinow (2012): 1060 "*Paean*"; Aeschin. *Emb.* 2.163; Aristoph. *Kn.* 1317–18. For prayer and sacrifice to Apollo for military purposes, see Hom. *Il.* 1.443–458.
42 Xen. *Ana.* 6.1.5.
43 Xen. *Hell.* 7.2.23.
44 Ath. *Dei.* 14.630–631. The poetry recited were verses from Tyrtaeus. See Bayliss (2017) for further reading.
45 Aeschin. *Emb.* 2.163; Aristoph. *Kn.* 1317–18; Aesch. *Pers.* 384–395.
46 Xen. *Ana.* 1.8.16–19, 4.3.31, 6.5.25–26, 6.5.29.
47 Rutherford (1994): 113; Pritchett (1971): 106; Plut. *Inst.* 16.
48 Plut. *Lyc.* 22.2–3.
49 A modern-day comparison of this continuing type of singing is evident in European football stadiums where repetition of question and answer songs are common. For example, "Everywhere we go, people ought to know, who we are and where we come from" or "Oh when the [insert team here] go marching in".
50 Bayliss (2017): 64–66.
51 Rutherford (1991): 1 and Strab. *Geo.* 9.3.10.
52 Harris & Platzner (2001): 205; Rutherford (1994): 114–115.
53 Rutherford (1995): 115.

54 Rutherford (1995): 115–116.
55 HH. 3.525.
56 HH. 3.300–304.
57 HH. 3.500; HH.3.517–518.
58 HH. 8.1–6.
59 HH. 8.15–17.
60 Plut. *Numa*. 13.
61 The Salii will be elaborated upon later in this chapter.
62 Plutarch used Marcus Terentius Varro and his *De Lingua Latina* here. Varro. *DLL*. 6.49.
63 Republican Roman history is admittedly vast and there are clear divisions between the early and late republican periods. For the purpose of consistency and succinctness of argument, be mindful of the definition for the "Republican" Roman military forces as mentioned in the opening paragraph of this chapter.
64 The earliest literary record surviving that refers to these traditional Roman battle expressions can be found in the works of Polybius (2nd century BC), Livy and Dionysius of Halicarnassus (1st centuries BC–AD).
65 Varro. *DLL*. 5.73, 7.49; Lloyd-Morgan (1996): 125–126; Keith (2002): 110; Wiseman (1982): 58–59; Williams (1965): 252; Dusanic (2003): 91.
66 Lloyd-Morgan (1996): 125–126.
67 Varro. *DLL*. 5.73, 7.49. Varro's etymology is probably not correct, though it may tell us something about Roman attitudes in the second century BC.
68 Dusanic (2003): 91.
69 Livy. 8.9.4–14.
70 Livy. 8.9.4–14.
71 Livy. 10.19.17–22.
72 Livy. 10.19.21.
73 Pythian Apollo, Jupiter Feretrius and Juno Sospita are some specific examples.
74 D. H. *Ant* 2.18.
75 D. H. *Ant*. 2.2.
76 D. H. *Ant*. 2.48.
77 Dio Cass. 54.8.
78 Dio Cass. 55.10.
79 Amm. 24.6.17.
80 Speidel (1984): 353–358.
81 Speidel (1984): 354.
82 Speidel (1984): 355.
83 Speidel (1984): 357.
84 Speidel (1984): 357.
85 Caes. *B. Afr*. 16.
86 Tac. *Hist*. 5.16.
87 Speidel (1984): 358.
88 Amm. 21.14.5.
89 D. H. *Ant*. 2.70–71; Varro. *DLL* 5.85; Plut. *Numa*. 13.
90 D. H. *Ant*. 2.70–71.
91 For further information regarding the Curetes, see Hom. *Il*. 9.529ff; Strab. *Geog*. 10.3.1 and D. H. *Ant*. 1.17.
92 Hempl, George. *"The Salian Hymn to Janus"* Transactions and Proceedings of the American Philological Association, Vol. 31 (1900), pp. 182–188.
93 Varro. *DLL*. 7.26–27.
94 Polyb. 15.12.8.
95 D. H. *Ant*. 8.66.2.
96 D. H. *Ant*. 8.66.2.

 97 D. H. *Ant.* 9.70. See Livy 10.40.12 for another example of Roman clashing arms and raising a "cheer".
 98 Plut. *Ant*. 39.4.
 99 Cowan (2007).
100 Cowan (2007): 115–116.
101 Cowan (2007): 114–115, 117.
102 Cowan (2007): 114.
103 Cowan (2007): 114–115, 117.
104 Rance (2015): 1.
105 Maur. *Strat*. 7.16.
106 Maur. *Strat*. 12B.16, 12B.24.
107 Maur. *Strat*. 2.18. Sub-heading "The Battle Cry Sometimes Used".
108 Maur. *Strat*. 3.15. Under sub-heading: "Instructions for the Troops in the Second Line".

7 Socio-political and military identity

Ancient military forces commonly presented themselves on the battlefield with the intent to express to the enemy and reaffirm unto themselves their socio-political identity. Armies undertook this to disassociate their force from the enemy; to generate otherness and difference from the enemy; and to display their force through compliance of what their collective group represented. Associating an army with a powerful element – symbol, custom or name of an influential leader – with a socio-political ideology aimed to inspire and instil a sense of superiority over the enemy. Armies expressed their identity on the battlefield in a variety of ways which included the adoption of an intentional uniformed appearance, the veneration of military equipment such as standards and trumpets, the use of sonic and visual techniques such as painted images and designs on the front of shields, and singing or chanting words/tunes that were associated with the army's socio-political background.

Appearance

Investigation into the way ancient military forces presented themselves on the battlefield reveals significant information regarding their socio-political and military identity. The intentional sonic and visual methods adopted by armies to communicate to the enemy and impress upon their own military force their identity is another feature of the battle expression. The manner that units within a military force and whole armies comported themselves in the lead up to the battle is incorporated within this paradigm. Armies from the Graeco-Roman world universally presented themselves according to an intentional image that sought to intimidate their opponents and inspire their own military units, often along cultural lines.[1] The adoption of uniformed military dress, military equipment and personal presentation, such as grooming, adopted by the rank and file embraces the notion of appearance. The psychological ramifications that an army's appearance had on the enemy could directly influence the outcome of a battle.[2]

Herodotus refers to the appearance of Arabian and Ethiopian warriors in terms of their clothing:

> The Arabians wore mantles girded up, and carried at their right side long bows curving backwards. The Ethiopians were wrapped in skins of leopards

DOI: 10.4324/9781003280439-7

and lions, and carried bows made of palmwood strips, no less than four cubits long, and short arrows pointed not with iron but with a sharpened stone that they use to carve seals; furthermore, they had spears pointed with a gazelle's horn sharpened like a lance, and also studded clubs. When they went into battle they painted half their bodies with gypsum and the other half with vermilion [red and white].[3]

Despite the functionality of the Arabian cloaks being worn on their right side, presumably for the purpose of utilising their bows, the appearance of this attire was deemed worthy of record by Herodotus for his audience, due to its peculiarity. The wearing of native, predatory animal skins (leopard and lion) coupled with the painting of the body using bright colours ensured Ethiopian warriors stood out on the battlefield. Whether there was any cultural meaning associated with the artistic style warriors painted their bodies or the colours chosen remains to be seen. What is evident in this passage are the clear attempts made by Ethiopian warriors to be seen on the battlefield and to display themselves in association with predatory animals. Perhaps the warriors wore the skins of animals that they had hunted and killed, or perhaps the physical capabilities attributed to the animals, whose skins were worn, aimed to reflect the warrior in some religious, military or cultural capacity.

Spartan hoplites, too, used clothing and grooming very particularly to present themselves on the battlefield.

In the equipment that he [Lycurgus] devised for the troops in battle he included a red cloak, because he believed this garment to have least resemblance to women's clothing and to be most suitable for war, and a brass shield, because it is very soon polished and tarnishes very slowly. He also permitted men who were past their first youth to wear long hair, believing that it would make them look taller, more dignified and more terrifying.[4]

Spartan armies customarily adopted an image of preparedness and satisfaction in the lead up to the battle.[5]

Once the enemy can see what is happening, a she-goat is sacrificed, and the law is that all the pipers present should play and every Spartan wear a garland; an order to polish weapons is also given. Young men may enter battle with their hair groomed . . . and with a joyful, distinguished appearance.[6]

Alexander the Great intentionally wore a helmet into battle to mark himself as visually distinct in comparison to other troops in his army. According to Plutarch, the shield that Alexander carried into battle, coupled with his helmet crest that held two large and bright white feathers on either side, illuminated him on the battlefield.[7] While this appearance served to highlight where Alexander was located and to permit his movements during battle to be tracked,[8] as often as not allowing the enemy to target him on the battlefield, much of the time the sight of Alexander leading his army from the front in the midst of battle both inspired enemy to dread, due to the courage displayed fighting in the thick of the enemy, the splendour of his armour and weapons, and the enthusiasm exhibited by himself and the men that surrounded him to engage with the enemy. This compelled his men to greater exploits.[9]

The uniformed appearance of the Roman army was an intentional image designed to impose fear over the enemy and familiarity amongst friendly troops. During the republic, plumes on soldiers' helmets served to intentionally heighten their appearance on the battlefield:

> Finally, the hastate wear as an ornament a plume of three purple or black feathers standing upright about a foot and a half in height. These are placed on the helmet, and the general effect combined with the rest of the armour is to make each man look about twice his real height, and gives him an appearance that strikes terror into the enemy.[10]

Roman military handbooks specifically encouraged the creation of a set image for Roman armies. The direction to maintain the cleanliness of weapons served to create an image of splendour that was deemed to strike terror into the enemy on the battlefield.[11] The purpose of Roman officers in covering their helmets with skins of wild, predatory animals was an intentional method to inspire fear in the enemy, but also to be more identifiable on the battlefield amongst friendly troops.[12] Roman military high commanders were fully aware that ethnically foreign enemies may appear different on the battlefield. That appearance could lead to the unsettling of the army. To stamp out the potential for Roman armies to succumb to fear from the peculiarities of other cultural groups, Roman commanders were encouraged to familiarise their armies with the appearance of the enemy before engaging in battle with them. Methods adopted by the Roman high command that aimed to reduce the negative impacts that could develop when a Roman army confronted an unknown force on the battlefield included: forming the army up in battle array in the presence of a hostile force while on campaign; through imitating the customs and appearance of a culturally diverse enemy, in a controlled training scenario (with friendly units playing the role of the enemy).[13]

> It is natural for the men in general to be affected with some sensations of fear at the beginning of an engagement, but there are without doubt some of a more timorous disposition who are disordered by the very sight of the enemy. To diminish these apprehensions before you venture on action, draw up your army frequently in order of battle in some safe situation, so that your men may be accustomed to the sight and appearance of the enemy. . . .
>
> Thus, they will become acquainted with their customs, arms and horses. And the objects with which we were once familiarized are no longer capable of inspiring us with terror.[14]

While distinct from Graeco-Roman military display, the appearance of Celt/ Germanic warriors shared the common purpose of invoking fear in the enemy yet being identifiable to friendly troops. Livy claims that Gallic warriors intentionally designed their battlefield appearance to strike terror into their enemies.[15]

> Tall bodies, long reddish hair, huge shields, very long swords; in addition, songs as they go into battle and yells and leapings and the dreadful din of arms as they clash shields according to some ancestral custom – all these are deliberately used to terrify their foes.[16]

Certain mosaics and sculptures preserve the desired image of different cultural groups on the battlefield, highlighting the cultural divergence of military styles, weapons and uniforms found within the Graeco-Roman Mediterranean world. Despite differences, these depictions reflect the pattern identified in literary descriptions, namely, despite variation in a presentation prior to engagement, there existed a desire to create an explicit battlefield appearance intended to impact in very particular ways on one's own forces and on the enemy.

The Alexander Mosaic highlights the intentional image created by different cultural groups present on the battlefield. The disparity seen between the weapons wielded by both armies, and the difference in military uniform and armour creates distinctive Macedonian and Persian images of identification. These identifiers served the purpose of distinguishing friend from foe in the heat of battle but also to present the enemy with a spectacle that communicated their military culture and origins. The mosaic found in a house in Pompeii was originally a painting from the Hellenistic world produced by painters that aimed to commemorate the military achievements of Alexander to an audience who would have been familiar with the military image in the mosaic. The Macedonian force is presented as unique compared to the Persian fighters as the infantry carry the *sarissa* and the Companion cavalry have their stylised metal helmet. The Persians, on the other hand, are presented with a cultural head piece made from textile fabric.

The sculptures found on Trajan's column in Rome serve to elucidate intentional military appearances adopted as a battle expression by Roman, with integrated Germanic troops, and Dacian military forces. The sculptural figures on this column offer clear distinctions in appearance between different cultural groups in a military context yet reaffirm how military forces used the display to present themselves on the battlefield in specific ways. As evident on Trajan's column Roman soldiers are depicted as being clad in helmet and breastplate, carrying *scuta*, *pila* and *gladii*. In stark contrast, Germanic warriors, serving within the ranks of the Roman army, are presented as shirtless, without armour or helmet, with facial hair and wielding a wooden club. Dacian warriors have portrayed bearing shields but otherwise without armour, sporting long hair and beards and wearing pants with a tunic or cloak. The military appearance of each culture is different in the way of weapons used and clothing/armour worn. Despite their difference in appearance, the common battle expression undertaken was to appear in a specific way that reflected native military culture to intimidate enemies and be identifiable amongst friendly troops on the battlefield.

Conversely, aside from intimidatory visual displays on the battlefield, armies could utilise massed sound to accentuate their appearance. To compensate for a smaller numerical force, or should the military high command feel that their army's low confidence level may render their battle expression audibly weaker to the enemy's, camp followers could be employed to strengthen an army's sonic and visual appearance. According to Livy:

> Marcus Marcellus on one occasion, fearing that a feeble battle cry would reveal the small number of his forces, commanded that sutlers, servants, and camp-followers of every sort should join in the cry. He thus threw the enemy into panic by giving the appearance of having a large army.[17]

The supplementation of non-combatants into the fighting ranks of an army, for the purpose of enhancing the sound and appearance of a battle expression, demonstrates the military importance of the practice.

Plutarch's biography of Gaius Marius describes the battle expression of the Celtic tribe "Ambrones". Plutarch states that the Ambrones' military force was hideous to look at, due to menacing gestures they made to the Romans and the size of their army. The noises they made from their speech and the bestial sounds they mimicked were unlike anything the Romans had witnessed before.[18] Marius forced his terrified Roman troops to avoid engaging the Celts until they had grown accustomed to their manner. After a short period of time, the Romans grew to fear the enemy less.[19] In time, the Romans took to the field against the Ambrones force. Before the battle commenced, the battle expression of the Ambrones was undertaken and was described as traditional, rehearsed and reflective of natural noise and power typical of Celtic/Germanic culture. Despite the Ambrones being heavily intoxicated and gorged with food, the warriors were not disorderly with their performance of a traditional battle expression. This took the form of rhythmic clashing of weapons together, massed leaping into the air simultaneously and shouting in unison their tribal name numerous times, "Ambrones!, Ambrones!".[20]

The appearance of thousands of Celtic warriors displaying their weapons and clashing them together *en masse*, coupled with the rhythmic leaping into the air and the corresponding impact of these men landing back on the ground, would have made this battle expression an awe-inspiring spectacle. Plutarch's reference suggests that this was a performance familiar to this cultural group as he refers to a military unit fighting on the Roman side, the Ligurians, who also shared ancestral lineage to the Ambrones. Plutarch claims that the Ligurians, too, began performing the same battle expression as the enemy. The Ambrones and Ligurians audibly duelled with each other, in an attempt to gain atmospheric superiority over the battlefield by creating the most amount of noise through rhythmic jumping, clashing of arms and chanting "Ambrones!, Ambrones!". The noise of the battle expression swung from one side of the battlefield to the other as each group took turns to perform, which served to heighten the fighting spirit of the combatants.[21] The atmosphere created by this exchange of noise would have been impressive. Plutarch's reference suggests that Celtic/Germanic battle expression types were based on proclaiming tribal ancestral origin.

Consistent with other cultures from around the ancient Mediterranean, Celtic and Germanic military forces had their own traditional and compulsory battle expression types that were performed before, during and/or after the battle.[22] The difference between Celtic and Germanic battle expressions from their ancient Greek or Roman equivalents is that the Celts and Germans continued their practice into the dark age and the early middle age by people from the same culture. The Celtic/Germanic tribes that were not completely conquered by Rome, along with the Germanic tribes that flooded south to occupy, or raid, the lands of the former western Roman empire during the dark ages continued to perform battle expression types that were similar, if not identical, versions of their ancient predecessors. Surviving Celtic/Germanic legends, sagas and histories from the middle ages record the battle expression of Irish, Scottish, Norman, Saxon and Frankish military forces. These sources give insight into what the nature of ancient battle expression performed by earlier Celtic/ Germanic military forces would have been like. These sources allow for a greater understanding of Celtic/Germanic battle expression to be obtained, in comparison to other ancient Mediterranean cultures, as these practices did not fade away with the fall of Hellenistic and Roman influence around the Mediterranean world.

The Irish mythological tale of Cuchulain can be used as an example to demonstrate the nature of Celtic/Germanic warrior transformation into animal-like warriors before battle. Celtic/Germanic warriors could adopt animalistic traits that were significant to their warrior style.[23] The transformation of human warrior into animal warrior (bear, stag, wolf) may have involved taking on the physical appearance of the so-called "warp-spasm" of Cuchulain. Preluding to the outbreak of the *berserk* rage, Celt/German warriors took on the appearance of physically transforming their bodies which is detailed below:

> The first warp-spasm seized Cúchulainn, and made him into a monstrous thing, hideous and shapeless, unheard of. His shanks and his joints, every knuckle and angle and organ from head to foot, shook like a tree in the flood or a reed in the stream. His body made a furious twist inside his skin, so that his feet and shins switched to the rear and his heels and calves switched to the front. . . . On his head the temple-sinews stretched to the nape of his neck, each mighty, immense, measureless knob as big as the head of a month-old child . . . he sucked one eye so deep into his head that a wild crane couldn't probe it onto his cheek out of the depths of his skull; the other eye fell out along his cheek. His mouth weirdly distorted: his cheek peeled back from his jaws until the gullet appeared, his lungs and his liver flapped in his mouth and throat, his lower jaw struck the upper a lion-killing blow, and fiery flakes large as a ram's fleece reached his mouth from his throat. . . . The hair of his head twisted like the tangle of a red thornbush stuck in a gap; if a royal apple tree with all its kingly fruit were shaken above him, scarce an apple would reach the ground but each would be spiked on a bristle of his hair as it stood up on his scalp with rage.[24]

According to the myth, Cuchulain was visited by his father, Lug, an Irish Celtic God, which may give insight into the religious significance of the transformation process of the warrior's body and mind.

The battle expression of the Gaelic clans of Ireland can be compared to those of the Celtic and Germanic military forces of antiquity.[25] Again, the Germanic and Celtic cultures in Ireland survived the fall of the Hellenistic and Roman Empires and were thus not destroyed but continued into the Dark Ages and Medieval periods. Literary evidence of the battle expressions of the dark age and medieval Celt and German military forces provide us with an echo of what ancient military practices were like. While many ancient Gallic, German and British tribes succumbed to the might of Rome and assimilated into the empire, or ceased to exist, Irish Celtic tribes were not subjugated in antiquity to Roman control. As a result, Irish tribes maintained their ancient military practices into the medieval age,[26] particularly the practice of united crying of statements and significant words which had direct links to their clan. The united cries of Irish clans have been compared to the practice of the ancient Picts and later English "Red Coat" soldiers who both used colour to distinguish themselves on the battlefield from their enemy.[27] The Irish clans did not, necessarily, integrate colour into their battle expression, however it was through their united cries of specific statements or words that differentiated themselves from their enemies on the battlefield.

Ammianus details a Persian battle expression performed during the siege of Amida. As the fighting raged, Ammianus claims that the hills surrounding the city echoed and re-echoed the sounds of the battle expression of both the Romans and the Persians:

> The hills re-echoed with the shouts which rose on either hand. Our men extolled the prowess of Constantius Caesar, 'lord of all things and of the world', while the Persians hailed Sapor as Saanshah and Peroz, titles which signify 'king of kings' and 'conqueror of war.[28]

While the hills around Amida accentuated the vocal component of both combatant forces' battle expressions, it should be clear that the volume of noise created by the unified voices of tens of thousands of soldiers on both sides would have been immense regardless of topography. According to Ammianus, both the Romans and the Persians extolled the prowess of their supreme leaders. The Romans honoured their emperor while the Persians glorified their king Sapor. According to Ammianus' description, it appears that both sides performed their battle expression due to custom, but also as a means to counter or drown out each other's vocal cries. The tune to which the Persians glorified their king is not made known by Ammianus, however, the lyrics are recorded. Ammianus states that the Persians hailed their king Sapor as "*Saanshah*" and "*Peroz*", which Ammianus translates as meaning "king of kings", for the former, and "conqueror of war", for the latter.[29] The title given to the Persian king, "king of kings", reinforces the concept of traditional Asian battle expression characteristics presented through Graeco-Roman historical works. That being said, Asian military forces were presented in the Graeco-Roman historical record as being effectively arrogant and psychologically egotistic towards their enemies, reflective in the title for their king.

Sparta

Plutarch's *Life of Lycurgus* and *Instituta Laconica* are useful sources when dealing with the Spartan battle expression. In these works, details are provided regarding Sparta's use of music and poetry in military contexts that highlight Spartan religious, socio-political and military customs. According to Plutarch:

> Their [Sparta's] very songs had a stimulus that roused the spirit and awoke enthusiastic and effectual effort; the style of them was simple and unaffected, and their themes were serious and edifying. They were for the most part praises of men who had died for Sparta, calling them blessed and happy; censure of men who had played the coward. . . . In short, if one studies the poetry of Sparta, of which some specimens were still extant in my time, and makes himself familiar with the marching songs which they used, to the accompaniment of the flute, when charging upon their foes. . . . For just before their battles, the king sacrificed to the Muses, reminding his warriors, as it would seem, of their training, and of the firm decisions they had made, in order that they might be prompt to face the dread issue, and might perform such martial deeds as would be worthy of some record.[30]
>
> And when at last they were drawn up in battle array and the enemy was at hand, the king sacrificed the customary she-goat, commanded all the warriors to set garlands upon their heads, and ordered the pipers to pipe the strains of the hymn to Castor; then he himself led off in a marching paian, and it was a sight equally grand and terrifying when they marched in step with the rhythm of the flute, without any gap in their line of battle, and with no confusion in their souls, but calmly and cheerfully moving with the strains of their hymn into the deadly fight. Neither fear nor excessive fury is likely to possess men so disposed, but rather a firm purpose full of hope and courage, believing as they do that Heaven is their ally.[31]

These references reveal that Spartan battle expression was a systematic process that held significant meaning. Spartan social custom is acknowledged in the praise of men who died on the battlefield and the condemnation of cowards. Religious sentiment is evident with the sacrifice to the Muses of a she-goat; the wearing of garlands on hoplites' heads; the tune to the hymn of Castor played on the flutes; the recital of the paean (presumably to Ares) and the hope and courage the army exhibited believing that the gods were with them. Spartan military training is revealed through references to rehearsed and predisposed processes that aimed to trigger specific action and mindset. For example, hoplites were exposed to marching songs in accompaniment with the flute as the army marched together to engage with the enemy. The sacrifice to the Muses geared the soldiers' mindset upon the training and effort required for the upcoming battle. The disciplined and methodological approach to battle the Spartans were described as displaying further supports the notion that battle expression was significant and meaningful. It was the battle expression that reminded soldiers of their training and orientated their psyche on the task at hand. The use of poetry in the above extracts refers to the impact that 7th-century BC elegiac poet, Tyrtaeus, had on Spartan military custom.[32] Plutarch claims that Tyrtaeus' poetry, essentially, resulted in Spartan soldiers craving death,[33] due to the enthusiasm it evoked within them.

The poetry of Tyrtaeus and the laws of Lycurgus inspired Spartan battle expression types.[34] According to Athenaeus, Spartan marching songs, *enoplia* (meaning under arms or martial rhythm), were sung when marching to battle. The lyrics of these songs consisted of the poetry of Tyrtaeus, which the Spartans recited from memory. According to Athenaeus, after the Spartans had defeated the Messenians through Tyrtaeus' leadership, the Spartan military initiated the custom of competitive singing. To ensure the lyrics of Tyrtaeus' poetry were remembered amongst the army while on the campaign, after meals were consumed and the paean was sung in thanksgiving Spartan warriors would compete over who could best sing Tyrtaeus' poetry. The competition would be presided over by the commander of the army where a piece of meat was awarded to the victor:

> The Spartans dedicate themselves to war, and their sons memorize their marching-songs, known as enoplia. So too the Spartans themselves recite Tyrtaeus' poems during their wars and move in time with them. Philochorus says that after the Spartans defeated the Messenians because of Tyrtaeus' generalship, they made it a custom during their campaigns that, after they have dinner and sing a paean, they take turns singing Tyrtaeus' poems; the polemarch judges among them and awards the winner a piece of meat as a prize.[35]

This reference highlights key features of the Spartan battle expression. The Spartan army used the poetry of Tyrtaeus customarily in the period before the battle commenced. The memory of Tyrtaeus' successful generalship over the Messenians, through the singing of his poetry, was used as a means for Spartan warriors to remember their hero of old and past military achievements of the Spartan state. The lyrics of Tyrtaeus' poetry aimed to instil within the reciter and the witness the qualities of Spartan soldiers and the doctrines of the Spartan state.

Singing *en masse* the poetry of Tyrtaeus aimed to instil steadfastness and unity within the army in the lead up to the battle. As the war songs of Tyrtaeus were used as marching songs, this may suggest that the primary aim of singing them was not to intimidate the enemy, but rather to inspire the men singing them. As a summary of the different war songs, the lyrics were designed to remove fear from the minds of the soldiers about to enter into battle and remind the soldiers of their role and obligations to their society.[36] The main recurring ideas that are found within Tyrtaeus' poetry are threefold; the qualities a Spartan warrior should display; the qualities a Spartan warrior should not exhibit and the significance of Sparta's homeland.

According to Tyrtaeus, excellence was gauged not by athleticism, strength, speed, personal appearance, wealth or singing ability, but by a man's ability to face the bloody slaughter of battle standing beside his compatriot.[37] Men who were able to achieve this were of more benefit to their city and people than those who could not. These men brought glory and renown to their homeland and their family was honoured and remembered because of the sacrifice made. The poetry of Tyrtaeus aimed to encourage Spartan citizens to strive for the qualities of respect, honour and excellence. These were characteristics that Spartan citizens aimed to achieve through their actions on the battlefield. In order to receive respect and honour, Spartan warriors had to fight and die courageously in the frontline facing the enemy. While fighting in battle, Spartan warriors were advised to demonstrate their skill in arms, valour, steadfastness and ferocity.

The concept of death in battle is portrayed in Tyrtaeus' war songs as being an inescapable feature of life that should not be feared; however, should a man die in his prime of life, gloriously fighting in the protection of his family and his homeland without shirking, then, respect and honour would be guaranteed. The Spartan sentiment towards achieving respect and honour by dying on the battlefield is reminiscent, and comparable, to the concept of martyrdom for religion and state, particularly demonstrated in literature during the Iran–Iraq war during the 1980s.[38] During this war, martyrdom became one of the most prominent themes in the official state discourses of both sides.[39] It attributed specific religious and nationalist meanings to those that were killed that continue to resonate in both countries' public discourse and collective memory of the war.[40] Wartime literature written about the conflict in Iraq glorified martyrdom and presented life as inferior to death.[41] For soldiers to die as martyrs was viewed as an honourable achievement that was envied by those who were unable to die in a similar fashion.[42] Families of martyrs were socially and financially rewarded inspiring soldiers to readily die in combat.[43] The enviable benefits of dying in battle at a young age found in Tyrtaeus' poetry and in the literature produced during the Iran–Iraq war include the glorification of the deceased amongst their people forevermore.

Tyrtaeus' war songs, contrastingly, discouraged Spartan warriors from exhibiting character flaws that would be considered shameful. A Spartan who abandoned his homeland in favour of vagrancy and a peaceful existence with his family was considered dishonourable.[44] Being an active member of Spartan society strengthened the bonds of fraternity and community. Should a man voluntarily abandon his society, particularly during times of war, Tyrtaeus presents this as shameful. However, should a man fight for Sparta the expectation was he never flee the battlefield, hide or display fear. Tyrtaeus notes that those who die in the midst of running, hiding or in a frightened state should be considered shamed.[45]

Tyrtaeus identifies the origins and experiences of the Spartan people as integral points of reference when inspiring Spartan warriors to fight and die for their homeland. Tyrtaeus' war songs reference the Spartan people being descended from Herakles. It appears the upholding of Herakles' courageous reputation and the honouring of ancestral lines was a significant feature in motivating Spartan warriors before battle and helped remove any apprehension Spartan hoplites may have had. Due to Zeus being the father of Herakles, the notion that Zeus favoured Sparta because of this is revealed in the poetry of Tyrtaeus, with the suggestion that Zeus gave the land to the Spartans because of their relationship with his son, Herakles.[46] Tyrtaeus presents Spartan society as being one that had endured a host of experiences that made it worthy of its citizen body to defend and sacrifice themselves for it. The acknowledgement of the Spartan state's familiarity with war, in all its forms, including victory, defeat and hardship, was used to inspire resolve amongst the fighting force prior to battle.[47]

Sparta's use of Tyrtaeus' war poetry in the lead up to the battle is undeniable.[48] Firstly, Tyrtaeus' war songs were recited on the military campaign before, during and after the battle. Secondly, Tyrtaeus composed marching songs and, finally, the Spartan army integrated Tyrtaeus' works into their pre-battle custom to enhance the levels of courage amongst the troops. The recital of Tyrtaeus' war poetry and war songs by hoplites reinforced the core socio-political values of Sparta within their military ranks. These core values are evident within the surviving extracts of Tyrtaeus' poetry, which are summarised neatly by Bayliss, "the imperative to fight bravely, the importance of collective responsibility, their own entitlement to rule the Peloponnese, and, perhaps most importantly, the inferiority of the helots".[49] There are only a small number of surviving Tyrtaeus fragments from antiquity. Given this, it has been argued that it was these verses that survived, particularly, that were overly consumed and used during classical times by Sparta's military would have taught Spartan youths the core Spartan values.[50] These few verses memorised by Spartan hoplites and rehearsed in contexts away from the battlefield would have made their rendition more effective in front of an enemy force due to the familiarity males had with the verses.

Tyrtaeus' war poetry and its affiliation with Sparta's military complies with the concept of battle expression. Tyrtaeus' poetry held significant meaning to Spartan hoplites and its recital was embedded within the sphere of military life. Scholars have not associated the military use of Tyrtaeus' poetry with the term war cry. This is due to the idea that Tyrtaeus' poetry was considered more than a battlefield custom but a key feature of Spartan military tradition. The term war cry does not allow for these ideas to be considered and gives weight to the argument that the notion of the war cry does not satisfy an ancient military phenomenon that held cultural meaning to the military force that used it.

Thucydides details a Spartan battle expression during the Peloponnesian War as reflective of the ideas that were presented in Tyrtaeus' poetry. During a battle narrative between the Argives, and their allies, against Sparta, Thucydides presents the Spartan army as being in full voice of encouragement for each soldier to be brave and to remember what they all knew so well. This reference relates to the notions inspired from Tyrtaeus' war songs.[51] Tyrtaeus' poetry reminded the Spartans of their duty to the state and urged courage, bravery and excellence in battle. Thucydides dubbed Sparta's war songs as πολεμικῶν νόμων – literally *war laws*. This term suggests that Tyrtaeus' poetry was well integrated within pre-battle custom.

The formulation of a stylised martial image that was impressive to behold was characteristic of Spartan battle expression.[52] This image was initiated by Lycurgus into Spartan military practice and consisted of wearing a red cloak and carrying a bronze shield. The purposeful inclusion of the red cloak for use in battle was deemed to be the most warlike of colours, devoid of peaceful connotations and more foreboding on the enemy.[53] Xenophon claims that Lycurgus chose the colour red as it was in contrast with female clothing. The adoption of bronze as the preferred metal for Spartan hoplite shields was, similarly, selected due to its impressive appearance once polished and practicality, as it was slow to tarnish. Spartan warriors, too, were permitted, and indeed encouraged,[54] to grow their hair long from adulthood for martial purposes. The Spartans believed that long, well-groomed hair would add to their commanding presence on the battlefield in the eyes of the enemy.[55]

Spartan awareness and purposeful creation of a warlike image in battle were legislated.[56] Xenophon claims that Spartan hoplites were required by Spartan law to regularly maintain gymnastic training with the intention of sculpting their bodies to appear impressive in their own eyes and superior to their enemies. The notion and practice of creating an image of superiority in front of the enemy are further detailed by Xenophon who claims Spartan armies purposefully waited until the enemy could bear witness to their image of discipline and belligerence.[57]

Spartan hoplites habitually created a warlike appearance that embodied calmness and discipline prior to engagement with the enemy in battle. This stylised appearance aimed to collectively focus the hoplites for battle and to overawe the enemy. The image is described in the opening sequences of the battle between the Argives against Sparta during the Peloponnesian War. The Argives are detailed as advancing with haste and fury, while the Spartans advanced slowly and in step to the sound of many flute players playing a particular tune.[58] Plutarch claims that the tune in question was the hymn to Castor.[59] Thucydides mentions that Sparta's slow and disciplined advance was customary of their military practice, so that the battle line may keep its shape and prevent the development of gaps, which often resulted with a fast-paced advance. Plutarch echoes this idea claiming the Spartan slow disciplined advance into battle marching in step to the tune of the flute with no gap, or confusion, in their line generated calmness and clarity of purpose in the Spartan army, but shock and awe amongst the on looking enemy.[60]

Sparta's disciplined battle expression was evident at the battle of Plataea during the Persian Wars.[61] Herodotus provides insight into the reaction of Spartan troops to adverse conditions that had developed during the opening stages of the battle. The Spartans were required to complete religious obligations, namely receiving favourable omens from pre-battle sacrifice. The favourable omens sought after were not forthcoming. While the army waited for divine favour, the Persians opposite them launched their attack on the Spartan frontline using archers. The Spartan army, despite sustaining many casualties, remained resolute in waiting for a successful sacrifice before being ordered to advance against the enemy. Herodotus claims that favourable omens were eventually received, and the order was given to engage with the enemy. The discipline demonstrated by the Spartans highlights the enduring image they wished to create for themselves as devout religious adherents, who were willing to die for their religious observance, making them worthy recipients of the god's favour. The utter contempt shown by the Sparta's army towards the enemy and their attack reveals the courage, disciplined training and resolve contained in Spartan pre-battle tradition.

For the Spartans, the demonstration of discipline and marching in step to the sound of the flute in the lead up to battle was a determinant of courage. In Plutarch's *Life of Agesilaus*, Agesilaus responds to the question: why the Spartans marched into battle to the sound of flutes? Agesilaus is claimed to have retorted, that marching in step to flutes distinguishes who is brave and who is a coward.[62] Spartan hoplites who were able to march in time to music were deemed to be collectively focused on the battle at hand rather than being preoccupied with thoughts of fear or personal safety. Plutarch's reference supports the notion that the image of bravery and sacrifice of self for the state as found in Tyrtaeus' war songs and poetry were customary of Spartan battle expression.

The Spartan battle expression was a process. It involved the army marching to battle and waiting for the enemy while singing their war songs inspired from Tyrtaeus' poetry. This instilled resolve and courage within the army to honour themselves and their polis. Once the enemy was within sight, a she-goat would be sacrificed, in conjunction with flute players striking up a tune, as each Spartan wore a garland in their hair. The respectful adherence to religious ritual demonstrated piety to the enemy. While the omens were interpreted, Spartan hoplites would polish their shields and comb their hair to create an image of discipline and commitment in front of the enemy. Once the hoplites were prepared the appearance of massed polished bronze shields, red cloaks and long hair from underneath helmets would have confronted the enemy across the battlefield. Upon receiving favourable omens from the sacrifice, the Spartan army would begin to march in step to the sound of flutes in a disciplined formation, a formidable sight.

Spartan battle expression demonstrates unique characteristics not found within other ancient Greek military forces. Lycurgus' refinement of military practice, such as the adoption of red cloak, bronze shield and long hair, suggests military insight and contemplation. The use of flutes and songs to recall Spartan social and military customs highlights nationalistic motivations. Through legislating features of battle expression, namely the growth of hair for male Spartan citizens, the Spartan government endorsed this tradition. The offering of blood libation, animal sacrifice and augury before battle suggests the close affiliation Spartan battle expression had with religious belief structure, particularly the recital of religious hymns and the playing of flutes.

Greek shield *semata*

The designs and motifs (*semata*) displayed upon Greek hoplite shields were a method adopted by Greek armies to express their identity on the battlefield. The emblems and designs used by Greek armies are presented in Graeco-Roman literature and iconographical sources such as vases. In battle, the exterior of hoplite shields faced the enemy across the battlefield. The shield designs served to express the socio-political identity of the army if the design was consistent throughout the whole phalanx or the identity of the individual hoplite. The earliest evidence for Greek shield devices comes from Late Geometric and Protoattic vases.[63] These early shield designs contained geometric patterns and animal motifs including gorgon's head, lion's head, bull's head, boar and flying birds and eagles. The origins of Greek *semata* on hoplite shields could be interpreted by modern students as having psychological or ritual origins, but the evidence would be difficult to find.[64] Lead votive figurines dated from the 6th century BC found in Sparta portray Spartan hoplite shields with geometric patterns and images of fierce animals and legendary figures.[65] For these early shield designs not to be viewed as a clear attempt to evoke intimidatory feeling upon the enemy across the battlefield, irrespective of individual designs on shields, seems strange. For fierce animals and legendary figures from Greek mythology, with violent and powerful connotations (such as gorgon) to be painted on the front of shields clearly demonstrates intent to instil a sense of awe upon the enemy in the recollection of the image on the shield. References to Greek *semata* in literary works that range from the 5th to 4th centuries BC present a context that suggests shield devices were, in fact, employed for psychological and ritual functions. The expression of uniformed identity was central to the use of *semata* on shields during this period.

The Greek military tradition of painting designs on the front of hoplite shields, to display to the enemy, for the purpose of identification and attempts to intimidate those across the battlefield, is evident in Graeco-Roman literature.[66] Theban battle expression consisted of promoting the club of Herakles on the shields of their hoplites. Xenophon's account of the battle of Mantinea in 362 BC[67] exemplifies the tradition of Theban hoplites displaying the club of Herakles on their shields before battle to display to the enemy. The club of Herakles served to identify Theban warriors from others on the battlefield, as well as to identify the polis of Thebes as having a special relationship with the Greek demi-god, Herakles. By displaying this symbol, Theban warriors aimed to replicate his strength on the battlefield and intimidate their opponents in the process. In Greek mythology, Herakles was the son of the god Zeus and mortal woman Alcmene and was famed throughout the Graeco-Roman world for his feats of strength, honour and victory. Theban military forces' association with Herakles in battle attests to their individuality compared to other Greek *poleis*. Greek city-states aimed to profess their historical origins, levels of piety and perceived military traits on the battlefield through clear visual demonstrations.

Literary evidence reveals that Spartan hoplites adopted a common *semata* on their shields that represented their military identity on the battlefield. According to ancient literary texts, the *lambdas* design was used by Spartan military forces during the 5th and 4th centuries BC. There is a reference from Eupolis' *Testimonia and Fragments* that mentions the lambdas as a symbol of fear.[68] This fear was aroused due to its military association with Spartan hoplite armies. During the 5th and 4th centuries BC, Spartan armies took on a uniform appearance with one key element being the letter lambda for *Lakedaimon* displayed on the front of their hoplite shields.[69] It is evident that other Greek states adopted the use of letters as a feature of uniform shield designs during this period too. Xenophon refers to a 4th century BC Spartan military force swapping shields with the Sicyonians which bear a sigma design.[70]

> But Pasimachus, the Lacedaemonian commander of horse, at the head of a few horsemen, when he saw the Sicyonians hard pressed, tied his horses to trees, took from the Sicyonians their shields, and advanced with a volunteer force against the Argives. The Argives, however, seeing the Sigmas upon the shields, did not fear these opponents at all, thinking that they were Sicyonians. Then, as the story goes, Pasimachus said: "By the twin gods, Argives, these Sigmas will deceive you," and came to close quarters with them; and fighting thus with a few against many he was slain, and likewise others of his party.

The appearance of the sigma shields in Xenophon's reference demonstrates the paradigm that is battle expression. The intentional uniformed appearance of a military force served identification purposes on the battlefield. The Sicyonian appearance in this instance did not inspire trepidation on the enemy as was intended the enemy did not view the sigma shields with any real threat. Despite this, the preparation levels and military thinking that would have been involved with the universal appearance of Greek hoplite armies, particularly during the 5th and 4th centuries BC, reflect the battle expression model which the term war cry fails to acknowledge.

Macedon

The kingdom of Macedon, likewise, used battle expression to differentiate itself from other Greek states on the battlefield. Arrian reveals that the Macedonian battle expression complied with the Greek cultural traits of discipline and piety, as mentioned earlier, however, embraced unique elements that served to represent Macedonian military individuality. The Macedonian spear, the *sarissa*, was developed during the military reformations of Philip II.[71] The *sarissa* gave the Macedonian phalanx a distinct advantage on the battlefield over its rivals, as it measured almost double the length of the spear used by the Greek hoplites and almost three times as long as the Persian spear. The intense training and discipline of the infantry with the *sarissa* under Philip II were continued into the reign of Philip's successor, Alexander.[72] The discipline and training of the Macedonian infantry with their *sarissa* inspired its use as a battle expression in the lead up to the battle.

Alexander's early reign saw him invade Illyria to crush insurrection there. During his actions to overcome the Taulantians, Alexander's force had found itself in a precarious situation of vulnerability. The Taulantians held a distinct military advantage by holding higher ground that restricted the movement of Alexander's forces. According to Arrian, Alexander ordered his heavy infantry, with their *sarissas*, to form together into a deep phalanx this formation would serve to accentuate the massed spears. From this position, Alexander ordered complete silence while the phalanx was put through a series of drills which displayed to the onlooking enemy the discipline, training and foreboding appearance of the Macedonian *sarissa*. Arrian states that the infantry were ordered to erect their spears, then lower them as if in attack formation. From this formation, Alexander ordered his men, in unison, to swing their spears around in various directions to demonstrate their superior training and effectiveness as a fighting force. This Macedonian battle expression intimidated and unnerved the Taulantians to such a degree that they withdrew from their advantageous position:

> Alexander drew up the main body of his infantry in mass formation 120 deep . . . with instructions to make no noise, and to obey orders smartly. Then he gave the order for the heavy infantry first to erect their spears, and afterwards, at the word of command, to lower the massed points for attack, swinging them, again at the word of command, now to the right, now to the left. The whole phalanx he then moved smartly forward, and, wheeling it this way and that, caused it to execute various intricate movements. Having thus put his troops with great rapidity through a number of different formations, he ordered his left to form a wedge and advanced to the attack. The enemy, already shaken by the smartness and discipline of these manoeuvres, abandoned their position on the lower slopes of the hills. . . . Alexander called on his men to raise the war-cry and clash their spears upon their shields, with the result that the din was altogether too much for the Taulantians.[73]

Arrian claims that the Macedonians raised a cry and clashed their spears against their shields. This is a different practice to any that has been mentioned before regarding Greek battle expression. Greek armies universally sang *paean* hymns in battle and, generally, there was a cry unleashed in honour of Ares ('Ενυαλίω ἀλαλάζοντας).[74] The Spartans recited extracts of Tyrtaeus' poetry and Greeks states, such as Thebes, displayed their identity through adorning their shields with symbols. The use of the *sarissa* by the Macedonian army to generate a roar of noise when clashed against a shield presents a unique aspect to the Greek battle expression. The Macedonian military forces of Alexander the Great employed the *sarissa* to intimidate the enemy in the lead up to the battle, through massed coordinated movements and by generating noise when struck against shields.

Polyaenus refers to a Macedonian battle expression that demonstrated the conviction of Alexander the Great's army in their aim to overthrow Darius III of Persia.[75] In the lead up to the battle of Issus in 333 BC, Polyaenus records Alexander's order to his troops to fall to their knees in prayer, before engaging the Persian battle line. Polyaenus claims the Macedonians undertook this gesture to appear reverent in the eyes of the enemy and, therefore, soften the hearts of the enemy and lure them into a false sense of security towards them. Darius is believed to have thought this was an act of supplication to him and that a battle was not going to be fought. At the sound of the trumpet blast the Macedonian army, as one, rose up from their knees and charged the enemy with such ferocity and conviction that the Persian centre collapsed under the weight of the attack. The characteristic Greek religious reverence presented in this reference reveals the spiritual dimension that Greek battle tradition took.

Archaeological discoveries of inscribed lead sling bullets may reveal aspects of Macedonian battle expression during the reign of Philip II. Inscribed lead sling bullets can be categorised into three classes based on the nature of the inscription found on them.[76] The first class contains exclamations such as "conquer", "take it", "woe", "blood" and "candy". The second-class names a settlement or a people. The third class, which is most interesting for Macedonian battle expressions, contains personal names of the generals of armies of which the slingers were members. At Olynthus, a settlement that Philip II of Macedon captured in ca.348 BC, numerous lead sling bullets have been unearthed which bear the inscribed name of King Philip.[77] Of 500 lead sling bullets discovered at Olynthus, 100 are inscribed and many of these bear the reference to King Philip II of Macedon.[78] The study of lead sling bullets and their inscriptions may give rise to a further understanding of battle expression within the Graeco-Roman world. In the case of Macedonian battle expression, the presence of King Philip II of Macedon's name on lead sling bullets suggests that the praising of the commander identified the socio-political tradition of Macedon. The action of engraving the name of the king onto a lead sling bullet and firing it into the enemy highlights significant meaning. There appears to be a genuine importance for slingers to identify themselves, their trade within the military (slinging) and their want for the enemy to literally feel that identity through injury, death or reading a misfired bullet by inscribing the name of their king and military commander on their tool of the trade. Whether ancient Macedonian warriors typically made a point to acknowledge the name of their king in battle, beyond the inscribed lead sling bullet, can only be suggested. What is interesting about the study of lead sling bullets is that it was important for soldiers within the army to express their socio-political origin by publicly acknowledging King Philip on weaponry.

Roman military triumphs

The study of the Roman military triumph may provide valuable insight into understanding the forms and significance of the battle expression based on socio-political identification. The military triumph should not be classified as a battle expression. Central to this understanding is that a battle that the triumphal procession may have commemorated ended days, weeks or many months before the celebration, often in a different part of the Roman world.[79] Additionally, the purpose of the battle expression was to unite by way of inspiring a military force, as well as, attempting to intimidate the enemy gathered on the field. With the battle well and truly completed and the location distant from Rome, this purpose no longer existed. However, the triumph does contain elements of the battle expression that was unique to the socio-political traditions of Rome and attached to their military culture. These customs helped forge Rome's military identity and transitioned onto the battlefield through the army's appearance. These include the archaic practices found within the triumph and the religious significance the triumph held to those involved which appear consistent with Graeco-Roman literary sources.

Modern study on the triumph suggests Etruscan and Hellenistic roots, which have been likened to the mythological Dionysian procession the Greek word *thriambos* directly stems from this idea and clearly has a religious dimension.[80] Archaeological evidence dating from the 6th century BC, particularly the site of where the temple of Jupiter Optimus Maximus on the Capitoline, holds the key to its foundation that is beyond recovery.[81] The origins of the military triumph are steeped in the distant past of Rome's history. Dionysius of Halicarnassus claims that Romulus celebrated a triumph. The triumph of Romulus, it appears, institutionalised what was to become standard Roman practice:

> He [Romulus] led his army home, carrying with him the spoils of those who had been slain in battle and the choicest part of the booty as an offering to the gods; and he offered many sacrifices besides. Romulus himself came last in the procession, clad in a purple robe and wearing a crown of laurel upon his head, and, that he might maintain the royal dignity, he rode in a chariot drawn by four horses. The rest of the army, both foot and horse, followed, ranged in their several divisions, praising the gods in songs of their country and extolling their general in improvised verses. . . . Such was the victorious procession, marked by the carrying of trophies and concluding with a sacrifice, which the Romans call a triumph, as it was first instituted by Romulus. But in our day the triumph has become a very costly and ostentatious pageant. . . . and it has departed in every respect from its ancient simplicity.[82]

This extract is highly useful in understanding the religious dimension of the battle expression. The connection between religious belief and military practice is evident. Ritual sacrifice and the praising of the gods in song encompasses the deep spiritual meaning evident in the battle expression concept. The singing of traditional and improvised songs during triumphal processions demonstrates that the Roman military were accustomed to such practices and this may well have been a custom on the battlefield.

Roman deities invoked during triumphal processions were also the deities worshipped on the battlefield in battle expression. Oaths made to specific gods on the battlefield were often fulfilled at the culmination of the triumphal procession.[83] The offering of spoils to the gods as a covenant for victory was satisfied during the triumph, suggesting a link to the practices on the battlefield. During Marcellus' triumph, the soldiers honoured Jupiter Feretrius in song, this may not be surprising considering all triumphs culminated at the temple of Jupiter on the Capitoline.[84] However, the oath taken by Marcellus on the battlefield was the dedication of spolia opima to Jupiter Feretrius. The fulfilment of battlefield oaths could be undertaken during a military triumph. That certain Roman gods could be used as a source of inspiration on the battlefield is evident in the trophies of war that Sulla offered up to the gods at the conclusion of his triumph to Mars,

Victory and Venus.[85] According to Plutarch, Sulla claimed that success in war was due just as much to the gods and good fortune than good generalship and the strength of the army. To gain the favour of divine benefaction, the Romans incorporated religion into military contexts to supplement their military might on the battlefield.

The reference to songs, performed by victorious soldiers in a military triumph, highlights an institutionalised practice within the Roman army.[86] The singing *en masse* of traditional, as well as, improvised songs during a triumph suggests that these customs were inherent within the army and may well have been a feature on the battlefield. The hailing of a military commander as *imperator* on the battlefield after a victory is evident within Graeco-Roman literary sources and may have been the first step in being awarded a military triumph through the streets of Rome.[87] Varro explains that the soldiers who marched in the triumph characteristically chanted *Io triumphe* as they journeyed with their general through the city and up to the Capitol. This chant was claimed by Varro to have a connection to Bacchus.[88] This chant has been likened to an archaic hymn that may have aimed to evoke the archaic religious world – in an appeal for divine epiphany and support for the idea that the triumphing general in some way represented a god.[89] Livy refers to Roman soldiers parading through Rome invoking the spirit of Triumph by name, as well as singing their own praises and those of their general.[90] Not only were these songs in praise of the general but it was customary to ridicule and embarrass the triumphant general.[91]

Crude songs, *carmina incondita*, that soldiers sang as they triumphed are claimed to have been created by the soldiers for the purpose of performing during the procession.[92] The singing directed at the general, part in praise, part in ribaldry has usually been explained by reference to the deepest prehistory and primitive meaning of the ceremony.[93] The cultural humour evident within these songs can be used as examples in understanding the types of battle expressions that could be employed on the battlefield. Through the study of Roman soldiers' songs performed during a triumph, important understanding about battle expression may be gained. For instance, soldiers could sing both traditional songs, which were religious in nature and purpose, and the *carmina incondita* highlights spontaneity and imagination on the part of the soldiers, which were generally not religious in nature. These crude songs aimed to embarrass the military high command and bring humour to the spectacle for the enjoyment of the soldiers and, no doubt, the audience. The participation of the rank and file in singing these tunes would have created an awe-inspiring atmosphere and would have reflected the collective spirit of the military force. The battle expression was similar in type; traditional and spontaneous, and purpose; generating an inspiring atmosphere to increase the collective spirit of the troops.

The *carmina incondita* were apotropaic songs that aimed to ward off the evil eye, reminiscent of the satiric rite of reversal, typical in Roman society.[94] Livy refers to the triumph of dictator Camillus in ca. 390 BC "between the rough jests uttered by the soldiers [*inconditos iaciunt*], was hailed in no unmeaning terms of praise as a Romulus and Father of his Country and a second Founder of the City".[95] Livy details that the triumph contained songs of triumph and customary jokes sung by the soldiers – *carmine triumphali et sollemnibus iocis.*[96] During Caesar's Gallic triumph, Suetonius records a song that his soldiers sung (*inter cetera carmina, qualia currum prosequentes ioculariter canunt*)[97] as they marched through the streets of Rome:

> The Gallic lands did Caesar master; Nicomedes mastered Caesar. Look! now Caesar rides in triumph, the one who mastered Gallic lands. Nicomedes does not triumph, the one who mastered Caesar.[98]

The combination of 11 key words makes for the lyrics of this song to be easily remembered. Despite the tune of the song being unknown, the repetition of key-words would have made it memorable. Another song is recorded by Suetonius from Caesar's Gallic triumph:

> Men of Rome, protect your wives; we are bringing in the bald adulterer. You Fucked away in Gaul the gold you borrowed here in Rome.[99]

Both the songs above focus on the sexual promiscuity and orientation of the commander-in-chief. The lyrics and nature of the soldiers' songs exemplify the cultural wit that went into their creation. It appears that ethnicity and culture served to unite soldiers into bonds of unity in the triumph and on the battlefield. The evidence of soldiers singing during triumphal processions reveals that this was a typical feature of Roman military life that extended onto the battlefield too. Military triumphs should be used as a comparative study to the battle expression, as important archaic religious and cultural practices evident in the triumph reflect similar battlefield customs. The worship of significant war gods through oaths and hymns, as well as, singing songs that were reflective of cultural wit are fundamental features of Roman military practice.

   The study on the Roman military triumph[100] is significant as it provides context for the archaic traditions of the Roman military. Studying the recorded chants/ songs of soldiers as they marched through the streets of Rome reveals that the Roman army was accustomed to prearranged, cohesive massed vocal noise. The connection between the army's singing and Roman religious ritual highlights the intimate relationship religion played in Roman military affairs. Ultimately, the study of the military triumph supports the argument that Roman cultural customs (often archaic) were used by the military and customarily integrated into military contexts to unite Roman soldiers along ethno-cultural lines on the battlefield. As a result, it is evident that massed vocal noise based on traditional religious sentiment was a common feature of Roman military life. Knowledge obtained from the study of the military triumph, used in conjunction with battle accounts from Graeco-Roman literary sources, demonstrates that Roman battle expression contained elements of the nature (massed vocal noise) and significance (socio-religious) of triumphal processions.

Military standards

Military standards and trumpets played an important role in Roman battle expression. The military standards (*signum*), including the golden Roman eagle (*aquila*), held tremendous socio-religious and military importance to the army.[101] Roman soldiers took a military oath to never desert their standards in battle as they were viewed as important religious paraphernalia comparable to statues of the gods.[102] The punishment for failing to fulfil the military oath to the standards involved death by way of decimation.[103] Military standards, such as the eagle, were used as religious inspiration for the army. Tacitus claims that Roman soldiers prayed to the standards and the gods of war.[104] Josephus, in his account of the Jewish War, refers to the Romans using their military standards as cult equipment central for post-battle ritual. In one instance, the Romans were recorded as erecting their military standards on top of a captured tower and while clapping their hands sang a song of victory, likened to a religious hymn.[105] On another occasion, after the capture of the Temple in Jerusalem, the Romans carried their military standards into the Temple complex and erected them before offering sacrifice to them.[106]

The significance of the military standards, particularly the eagle, to the army is detailed by Speidel in his *Roman Army Studies*.[107] The eagle's association with Jupiter is emphasised as being instrumental in its role within the military.[108] This idea is evident in the writings of Josephus, who claims the eagle, at the head of the Roman army, was viewed as the king and strongest of the birds, Tacitus confirms this notion claiming the eagle was referred to as the symbol of the true deities of the legions.[109] Those military standards held prominence within Roman military religious life is overwhelming. For example, chapels were erected in every Roman military camp to store the standards; watch was kept to guard the standards; men were assigned to carry the standards into battle; standard bearers were treated with great respect within the army;[110] on festive days the standards were subject to sacred rites, such as anointing rituals.[111] In the lead up to the battle, the standards were looked upon for omens that could express or alter the mood of the men.[112] According to Josephus, the eagle was seen as a symbol of successful conquest over all whom the army advanced against and symbolised the honour of the Roman military.[113]

The sight of military standards in the battle line had a rousing effect on the army in the lead up to the battle. The mere reference to the standards in the lead up to the battle from the high command galvanised the army to engage with the enemy:

> Then (with God's leave be it spoken) let us advance our triumphant eagles and victorious standards. The soldiers did not allow him to finish what he was saying, but gnashed and ground their teeth and showed their eagerness for battle by striking their spears and shields together, and besought him that they might be led against an enemy who was already in sight.[114]

Military standards had fixed roles and positions on the battlefield.[115] It was in these positions that standards influenced actions the army would undertake[116] such as the advance to engage the enemy. It was during the advance against the enemy that those in the army undertook battle expression forms, such as swearing oaths, invoking the gods, clashing weapons against shields *et al.* The respect afforded to the standards makes them a key element in understanding the Roman battle expression. This is demonstrated by the influence they had over the actions of the army, particularly in the opening stages of a battle.

According to Graeco-Roman authors, the effect military standards had on the enemy across the battlefield was profound. The image of gleaming standards inflicting fear and intimidation upon Rome's enemies in battle is evident in the writings of Livy and Ammianus Marcellinus, specifically. In Livy's account of Scipio's campaign against Hasdrubal, the Carthaginian general is presented as being taken aback along with his men by the sight of Rome's gleaming standards and multitude of men.[117] The image of the foreboding gleaming military standards reoccurs in Ammianus where Germanic tribal armies lose heart at the sight of the standards:

> And so, when the signal had been given by the trumpet and they began to engage at close quarters, the Germans stood amazed, terrified by the fearful sight of the gleaming standards.[118]

Ammianus claims that during a Saxon intrusion into Roman lands in Gaul, the generals Nannenus and Severus:

> So terrified and confused the arrogant barbarians before the struggle began, that they did not oppose him in strife, but, dazzled by the gleam of the standards and eagles, begged for pardon and peace.[119]

Evidence suggests that Roman armies used their military standards as a form of battle expression. As the references above reveal, Roman soldiers took inspiration and guidance from their eagles and other standards. The standards were deemed sacred objects in the religious life of the army and were venerated on and off the battlefield. The significance of the standards to Roman soldiers on the battlefield is obvious. Aside from the resolve that soldiers took from their standards, the enemies of Rome viewed the gleaming standards with fear and trepidation. Surely the Romans observed this on the battlefield and as such used their standards in another facet of intimidating and unnerving the enemy before battle. As Ammianus claims earlier, the standards of Rome's army could essentially win a battle without bloodshed. Perhaps this was the potential influence the army aimed to capitalise upon, to utilise the imposing appearance of their standards to gain victory without bloodshed.

Trumpets

The use of trumpets[120] in battle can be categorised as a Roman battle expression and an identifier of a Roman army's appearance, compared to other military forces of the Graeco-Roman world. Roman military forces used trumpets to give orders on the battlefield and in camp.[121] Aside from orders issued via trumpet for action, such as duties, meals, military manoeuvres, striking camp, preparing for march, trumpets were used to initiate battle expression and were used as a form of battle expression. The trumpet served to unite and inspire Roman soldiers and instil fear within the enemy. Graeco-Roman authors refer to trumpets, associated with battle expression, from the early republic through to the late empire. The prevalence of references in the literary record, to inspire Roman soldiers and instil terror within the enemy on the battlefield, demonstrates the significant role these instruments played for the Roman army over a broad period.

Roman trumpet sounds on the battlefield were typically associated with the raising of a massed vocal shout/song/noise, connecting the trumpet to the battle expression.[122] The trumpets' initiation of vocal battle expression recurs in Livy's work,[123] for example: Marcellus ordered the trumpets to be sounded and a shout raised.[124] Dionysius of Halicarnassus, likewise, details the battle expression as being initiated by the trumpet blast followed by massed vocal noise.[125] Caesar, in his *Civil War*, reverses the process found in Livy and Dionysius as he roused his men through rhetoric to the point of vocal battle expression which resulted in the trumpets being sounded for battle:

> After this speech, when the soldiers were clamoring and blazing with enthusiasm for battle, he let the signal sound.[126]

Sallust in his history of the war against Jugurtha details a tactic Marius employed, who aimed to surprise his enemy by maintaining strict silence before unleashing their battle expression; the trumpet and massed vocal noise.[127] A type of battle expression consisted of all the trumpeters in the legion simultaneously sounding their instruments from which the troops undertook a massed vocal noise.[128] The combination of trumpet and massed vocal noise as a battle expression appears typical of Roman custom.[129] In the late empire, Roman armies continued with a similar, if not identical, battle expression. Ammianus Marcellinus refers to Roman trumpets initiating massed vocal cries during separate battle narratives.[130]

The lack of details regarding the tune played by the trumpets and the lyrics makes it difficult to piece together the complete battle expression that combined both sonic elements. Greater clarity can be made of the impact this type of battle expression had on the Romans and their enemy in the lead up to the battle. Polybius, in his summary of the battle of Zama in ca.202 BC, claims that the sound of the Roman battle expression caused the elephants, within the ranks of the Carthaginian army, to take fright and turn on friendly troops. When the trumpets and bugles sounded from all sides, some of the animals took fright and at once turned tail and rushed back upon the Numidians who had come up to help the Carthaginians.[131]

Graeco-Roman authors testify to the great and terrible noise that the trumpets and vocal noise created and the fear that overtook the enemy that bore witness to it.[132] Of note is the description made by Sallust when detailing the successful Marian surprise attack on Jugurtha's forces. After the enemy had excessively celebrated the night previously, thinking they had trapped Marius' Roman force. Marius waited to attack the enemy army in the morning he ordered strict silence upon his troops to ensure the noise of the battle expression would be optimal. As the order for trumpet and massed vocals was given, Sallust claims that:

> The Moors and Gaetulians, having been suddenly awakened by the strange and terrible sound, could not flee, arm themselves, or do or provide for anything at all; thus had terror, like a frenzy, seized everyone of them as a result of the clash of arms, the shouting, the lack of help, the charge of our men, and the confusion. In a word, they were all routed and put to flight, most of their arms and military standards were taken.[133]

The impact the sound of the trumpets had on the Romans was juxtaposed with the impact it had on enemy forces. Speidel claims that horns, particularly the *bucina*, were blown before battle so that the fighting fever would grip the men.[134] Appian claims that the trumpets aroused the soldiers with their inspiring blasts[135] Ammianus provides a similar description, citing the trumpet blasts that aroused the army.[136] It appears that confidence spread amongst the Roman army because of the trumpet battle expression.[137]

The combination of musical instruments and massed vocal noise on the battlefield reveals that high levels of rehearsal and coordination took place off the battlefield in preparation. The organisation that would need to take place for an effectual impact on the battlefield suggests sophistication, cohesion and coordination on the part of the army.

Massed vocal noise

Recent scholarly work acknowledges that Romans did use war cries to frighten the enemy and to raise morale amongst the army, and that Roman military forces traditionally practiced two types of war/battle cries.[138] Claims that a strict observance of intentional silence *en masse* was characteristic of the prelude to Roman battles against enemies.[139] The need to hear orders and maintain formation was vital in the opening stages of a military engagement. With that in mind, what modern scholars fail to acknowledge are the other types of battle expressions that were at the disposal of the Roman army. The lack of recognition for the cultural uniqueness and significance that the battle expression held for the soldiers on the battlefield is not directly present in modern work. Evidence from Cassius Dio for Rome's early imperial period and Vegetius and other writers for the later imperial period to support this.[140] Despite the focus on silence prior to battle, he does acknowledge that Roman military forces performed "battle cries" immediately prior to, or upon, engagement with the enemy at close quarters. This practice intensified the psychological impact on the enemy. As Vegetius notes:

> The war shout should not be begun till both armies have joined, for it is a mark of ignorance or cowardice to give it at a distance. The effect is much

greater on the enemy when they find themselves struck at the same instant with the horror of the noise and the points of the weapons.[141]

The acknowledgement that Roman armies traditionally employed a range of battlefield customs throughout its history, which diversified as their conquest of the Mediterranean world did, supports the battle expression concept. However, this does not detail the significance such actions had on the soldiers undertaking them and the socio-religious importance these traditions held within the Roman culture. Nor does this detail the military intent of these customs and the potential effects they could have on the enemy. When compared to the study of the Roman battle expression these ideas are limited to silence, clashing of weapons against shields, the *barritus* and Christian-inspired acclamation. There is no reference to other traditionally "Roman" customs, such as the use of brass instruments and military insignia, or no mention of Roman deities (except for the Christian God). The problems associated with the Roman civil war periods are overlooked and the adoption of non-Roman customs is restricted to the *barritus* and Christian invocations. Modern scholarly understanding of the war cry, that is, acknowledgement that there were battlefield practices adopted by specific cultural groups within the Graeco-Roman world, is limited in scope relating to the range of battle expression types and does not acknowledge the sophisticated nature or the significance of these customs held for military forces.

The lyrics of massed vocal noise are not detailed with great clarity in the literary sources. Whether these authors felt no obligation to write down the lyrics due to the audience's familiarity with the songs/chants/cries is unclear. Perhaps the sources of information that authors used, or in the case of eyewitness accounts the inability to hear, or remember, may be a likely explanation.[142] Whatever the case may be, it is clear that Romans did have a tradition of singing/chanting *en masse* in a battle context. The nature of these vocal arrangements was inspired by religious beliefs, socio-cultural factors or spontaneity given the situation at hand. The practice of Roman armies going into battle, during battle and/or after battle committing to massed vocal noise is prevalent in literary sources that range from the 1st century BC through to AD 4th century, and can, therefore, be categorised as a military phenomenon. Aside from Caesar's acknowledgement that Romans, indeed, practiced battle expression,[143] particularly vocal, evidence from Rome's early republic suggests Caesar's claims that the act of undertaking a battle expression was an ancient institution (*antiquitus institutum est*) is made apparent.

The reference to vocal battle expression is recorded by Graeco-Roman authors as being typically Roman alluding to a distinction between different military forces on the battlefield. Roman vocal battle expression is deemed unique to them, "the soldiers, raising their usual battle cry".[144] The translation associates the Roman battle expression as their own.[145] On the occasion when the early republican army, and their non-Roman enemy, both raise a battle expression it is stated that "they raised their war cries".[146] Whether the Roman military custom of clashing weapons against shield and the customs of the *Salii* connection can be, in a similar manner, applied to the vocal noise generated by the Roman army (especially during the early republic) remains to be seen. It could be inferred that the Salian hymns were transferred onto the battlefield and used by the military forces *veterem memoriam.*

Massed vocal noise was raised when the Romans forced the enemy to retreat. According to Plutarch's account of Marcellus' dedication of spolia opima to Jupiter Feretrius, the Roman exhortation to each other on the battlefield when in pursuit of a fleeing enemy to *feri* or "strike" was made known.[147] The fact that this reference was made in association with the origins of the epithet of Jupiter and not detailed in a battle narrative where the Romans had forced the enemy to flee is interesting. This may give weight to the idea that the audiences of Graeco-Roman historical works were acquainted with the vocal noises from the battlefield and had no need of detail during the battle narrative.

The impact that the vocal battle expression had on the army, and the enemy, of the early republic, attests to its significance. Reference to the army repeatedly performing a massed vocal battle expression in preparation for an attack portrays it as a morale-boosting exercise.[148] The repeated performance presents the audience with an image of soldiers attempting to generate camaraderie and resolve for the battle to come. The effectiveness of this undertaking created an atmosphere that was inspiring to the participants, due to the continual performance of the vocal noise.[149] The impact the battle expression had on the enemy was immense, as they withdrew from the battlefield before engagement. The avoidance of bloodshed through the implementation of massed vocal noise reveals that this form of battle expression could influence the outcome of a battle. This is evident during the opening stages of a battle between a Roman army and the Vosci, the ramification of an effective vocal battle expression left the Volscian confused and unable to remain on the battlefield.[150] The result was that the Volscian army fled to find refuge behind their settlement walls.

Military handbooks from AD 1st century highlight the nature, prevalence and purpose of massed vocal battle expressions. Onosander, in his *Strategikos*, recommends that generals send their army into battle making massed vocal noise (ἀλαλαγμῷ). In accompaniment with the clashing of weapons, the enemy would be adversely affected psychologically.[151] The collection of military principles and teachings detailed in Onosander's handbook was obtained from Roman practices.[152] Therefore, the referral to massed vocal noise, for the purpose of intimidating the enemy on the battlefield, was an established cultural tradition within the Roman military. It was recorded, by Onosander, for contemporary military commanders and for posterity, so that Roman generals would continue the practice.

Evidence reveals that Roman armies continued to utilise massed vocal noise to culturally identify themselves on the battlefield during the late empire. Ammianus Marcellinus records that the army lyrically praised the socio-political leader of the Roman world, the emperor, in a battle expression. Ammianus records the events surrounding the siege of Amida, AD 359, between Roman and Persian forces:

> The hills re-echoed from the shouts which rose from both sides. Our men praised the prowess of Constantius Caesar, 'lord of all things and of the

world', while the Persians hailed Sapor as Shahanshah and Peroz, titles which signify 'king of kings' and 'conqueror of war'.[153]

The battle expression detailed by Ammianus reveals that the army honoured individuals based on their leadership within the Roman world. It was through a massed harmonious chant that the army praised the superiority of their Emperor and commander-in-chief, although he was not present at the siege. Before the battle, the army acknowledged Constantius Caesar as the "lord of all things and of the world". The Persians, on the other hand, praised their commander-in-chief and king, "*Shahanshah*" and "*Peroz*". From the perspective of the audience, this exchange of battle expression between powerful military opponents seems quite competitive. Of significance is the back-and-forth (monomachy) nature of the battle expression vocalised by each army. This seems reminiscent of football supporters inside a stadium competing with rival supporters to outcheer the other for supremacy of the atmosphere and to gain a psychological edge over the opposition. Football supporter groups, too, impose their identity on the opposing supporter group and the team. The reminder of identity is used to enthuse the players of the supported team to perform to the best of their ability. Ammianus' reference contributes to the reader's understanding of the tension and rivalry that these forces felt towards one another prior to battle. However, for the rank-and-file present on the battlefield, the Latin battle expression of the Romans would have been difficult to understand for the common Persian soldier, likewise too, for those in the Roman force to understand the statements made by the Persian force.

Whether Ammianus devised this exchange between the military forces gathered to heighten the narrative of the battle sequence he was describing and place the battle in the perspective of international military and political importance, or whether this was an actual recording of events as they transpired is of interest. As will be detailed later, Ammianus' description of the "*Barritus*" battle expression, which is referred to in scholarly works that relate to "Roman war cries",[154] is understood to be reliable, and so there are grounds for this battle expression to be viewed in the same light. This is due to the author being present at this siege and being witness to the events. Of significance, in this reference is Ammianus' claim that the noise generated by both military forces was heightened by the terrain, namely the hills, which would have made the statements made by both sets of forces more definite and audible. This may be used to support the credibility of the claims made in this reference.

Ammianus' reference suggests that Roman and Persian forces typically praised their political leader as a battle expression, not to topically intimidate and undermine the enemy and their commander-in-chief whether present or not, but to reaffirm the military prowess of their own political leaders and social system. The manner that the opposing forces chanted or shouted *en masse* the statements about their leaders in the presence of foreign speakers were used to intimidate, psychologically, the enemy. Ammianus' statement reinforces the argument that Roman military forces incorporated typically cultural characteristics into their massed vocal battle expression. In this reference, the army extolled their political orientation through their acknowledgement and praise of the head of state, even though the emperor was not present at the siege. In the modern day, a comparison of this practice can be applied to "God save the Queen" in England.[155]

This battle expression reflects Roman culture embraced the practice of addressing social superiors (political/military/religious) with titles.[156] The title awarded to the emperor, proclaimed by the army, would have been reminiscent of the standard imperial court etiquette of Constantius.[157] Constantius was referred to on the battlefield with a title he, and other emperors before him would have been addressed with formally by their subjects. This notion can be compared to the honorific titles awarded to any socio-political leader of the modern world such as your majesty (or grace) of English monarchs and Mr President, for the President of the United States. The vocal battle expression from Ammianus suggests that the social customs of the Romans inspired their battle expression during the imperial age.

Archaeological evidence

Rome's absorption of Hellenistic slingers into their army brought with it the tradition of inscribing sling bullets with taunts and acclamations of identity.[158] Roman sling bullets have been excavated those bear inscriptions that contained the names of generals, legions and military units. Other versions contain images, such as lightning bolts and phalluses.[159] The inscriptions found on sling bullets may provide significant insight into the nature and purpose of Roman battle expression. Proclamation of identity (commanding general, legion and unit name) on sling bullets complies with the literary evidence that refers to Roman armies taking pride in their legion number/name expressed on the battlefield through massed vocal noise and standards. The images of lightning and phalluses on sling bullets may reveal elements of Roman religious tradition and the power contained within symbols representative of those beliefs. Lightning was synonymous with Jupiter, and the connection between Jupiter Feretrius and Roman military practice is evident in the literary source material. The phallus was symbolic of the Roman god Priapus but may well have been inscribed on sling bullets to signify the penetrative qualities that a sling bullet had like a phallus. For slingers within the Roman army to impose religious customs onto the enemy, through inscribed sling bullets – physically, correlates with what has already been revealed regarding the connection between battle expression and Roman religion.

The intimidatory nature of battle expression types is displayed through inscribed sling bullets. The study of inscribed sling bullets adds an extra dimension to the purpose and significance of the battle expression held militarily. The military intent of the battle expression and sling bullets was to impose Roman sentiment (identity, religious belief and cultural wit) over the enemy, for the purpose of inflicting serious psychological and physical injury on the enemy.[160] Roman sling bullets unearthed at Perusia, from the Perusine War ca. 41 BC, contain coarse sexual statements in the context of phallus imagery.[161] The wit and humour intended through these sling bullets, in particular, highlight unique features of Roman culture.[162] When compared with other examples from literary sources that encapsulate Roman cultural wit and humour, it becomes apparent that battle expression types were inspired from the cultural background. Roman armies insulted enemy forces using derogatory terms, often in a sexual context suggesting inferiority. During the Perusine War between the forces of Antony[163] and Octavian, both Roman armies utilised sling bullets to insult prominent individuals associated with the army, namely Fulvia and Octavian.[164] The culturally Roman nature of sling bullet inscriptions correlates with the battle expression and should be used as archaeological evidence to represent intentional attempts made by Roman armies to intimidate the enemy through uniquely Roman language and sentiment. The suggestion has been made that Roman sling bullets held religious significance with the god Priapus further highlights this argument.[165]

The connection between sling bullet inscriptions, military triumphs and Roman religious tradition in a military context is clear. When used in conjunction, Roman battle expression types can be categorised as containing ribald humour in a military context, intentional actions that reflect sophisticated organisation on the part of the military high command and are meaningful to the Roman religious tradition. Inscribed Roman sling bullets from the Perusine War contribute significantly to the study of the battle expression as it testifies to the culturally unique nature and significance of battle expression types.

Roman civil war periods and the integration of non-Roman forms of battle expression

The Roman battle expression during periods of civil war emphasised the unique cultural nature it took. Battles that took place during the Roman civil war periods resulted in the cessation of traditional battle expression forms, or the alteration and development into different practices. The lyrics acclaimed, the movement carried out, the clashing of weapons against shield and the blast of trumpets, customary of traditional Roman battle expression that aimed to culturally unite soldiers against non-Roman enemies in battle, proved problematic during civil war periods. The outbreak of civil war during the late republic and early imperial age pitted the Roman army against itself. The morale-boosting and intimidatory practices, that were effective against non-Roman enemies, did not have the same effect on these occasions.

Appian details the siege of Mutina from 43 BC between the Roman forces controlled by Octavian and Antony. In this battle scenario, the Roman armies opposing each other purposefully did not perform a battle expression. The reasoning behind this tactic, according to Appian, was that it would not serve to benefit either forces, due to the familiarity each side had with the Roman battle expression.

> Being veterans they raised no battle-cry, since they could not expect to terrify each other, nor in the engagement did they utter a sound, either as victors or vanquished.[166]

Of significance, the army assumed that they did not have a battle expression available that would have surprised or served to gain a psychological advantage over their compatriots. This may suggest the Romans primarily used culturally specific battle expression forms when fighting an enemy. The attempts to frighten the enemy through the manner of the battle expression seem obvious; however, the battle expression served to culturally unite the soldiers against non-Roman enemies. Roman civil war periods proved, as demonstrated through Appian, problematic in relation to the battle expression. Civil war periods created confusion and ineffectiveness for the battle expression as described by Tacitus:

> Throughout the night the battle raged in many forms, indecisive and fierce, destructive, first to one side, then to the other. Courage, strength, even the eye with its keenest sight, were of no avail. Both armies fought with the same weapons; the watch-word, continually asked, became known; the colours were confused together, as parties of combatants snatched them from the enemy, and hurried them in this or that direction.[167]

It was during the devastating AD 68–70 civil wars that the battle expression shifted to gain an advantage over a similar enemy. Stressing the identity and reputation of an individual legion was one method employed during the civil war to galvanise potentially unwilling armies to fight against compatriots. Tacitus details a key battle during the Batavian Rebellion (AD 69–70), legion fought against the legion and their commanders, familiar with the Roman military practice, aimed to evoke the spirit of individual divisions within their armies. Even though this rebellion aimed at liberating the tribes of Batavia from Roman rule and the military forces of Civilis contained many Germanic warriors, the Germanic tribes had long been enrolled in military service for Rome. The ethnicity of the Civilian army focused on the removal of any friendly association with Rome. Opposing Civilis was Cerialis' army (predominantly Roman in ethnicity). His focus on legion identity and honour while emphasising a stable empire was used to overcome this difficult military encounter:

> Cerialis recalled the ancient glories of the Roman name, their victories old and new; he urged them to destroy forever these treacherous and cowardly foes whom they had already beaten; it was vengeance rather than battle that was needed. "You have recently fought against superior numbers, and yet you routed the Germans, and their picked troops at that: those who survive carry terror in their hearts and wounds on their backs." He applied the proper spur to each of the legions, calling the Fourteenth the "Conquerors of Britain," reminding the Sixth that it was by their influence that Galba had been made emperor, and telling the Second that in the battle that day they would dedicate their new standards, and their new eagle. Then he rode toward the German army, and stretching out his hands begged these troops to recover their own river-bank and their camp at the expense of the enemy's blood. An enthusiastic shout arose from all.

Nor did Civilis form his lines in silence, but called on the place of battle to bear witness to his soldiers' bravery: he reminded the Germans and Batavians that they were standing on the field of glory, that they were trampling underfoot the bones and ashes of Roman legions. "Wherever the Roman turns his eyes," he cried, "captivity, disaster, and dire omens confront him. You must not be alarmed by the adverse result of your battle with the Treviri: there their very victory hampered the Germans, for they dropped their arms and filled their hands with booty: but everything since has gone favourably for us and against the Romans. Every provision has been made that a wise general should make: the fields are flooded, but we know them well; the marshes are fatal to our foes. Before you are the Rhine and the gods of Germany: engage under their divine favour, remembering your wives, parents, and fatherland: this day shall crown the glories of our sires or be counted the deepest disgrace by our descendants!" When the Germans had applauded these words with clashing arms and wild dancing according to their custom.[168]

Admittedly, the speeches recorded by Tacitus are fictional and inspired by his own literary tradition, however, the sentiment found within them may not necessarily be far from the truth. Of significance, in the speeches, by AD 69–70, the composition of the army was heavily influenced by non-Roman warriors. The emphasis on avenging Rome and legion identity, by Cerialis, suggests the composition of his forces may not have necessarily been ethnically Roman. The agenda to remove Roman association from Civilis' army, but instead focusing on German customs highlights the complex and racial mixture that the Roman army had already found itself during the early imperial period.

The competition and rivalry between the legions, driven by their honour, reputation and spirit/Genii may help explain the determination of Roman armies fighting against each other during the civil war between Otho and Vitellius in AD 69. Tacitus makes a clear distinction between legions in his battle narrative between the forces of Otho and Vitellius. The specific details provided regarding the successes and failures of legions suggest that the types of battle expression adopted by these armies were inspired by unit/legion identity:

On that of Otho was the 1st, called Adjutrix, which had never before been brought into the field, but was high-spirited, and eager to gain its first triumph. The men of the 1st, overthrowing the foremost ranks of the 21st, carried off the eagle. The 21st, infuriated by this loss, not only repulsed the 1st, and slew the legate, Orfidius Benignus, but captured many colours and standards from the enemy.[169]

This extract details the achievements of individual legions within the Roman armies that fought against each other in the recorded battle. The identification of the Roman legions as the "1st" called "*Adjutrix*" and the "21st" and not simply a Roman legion relates to the problem Roman civil war periods had on the identity of the Roman army. This extract suggests that the focus on individual legion

names and numbers within the Roman army took on a different form of importance for group identity during civil war periods. To generate the feeling of otherness from the enemies who were essentially the same, this extract suggests that
Roman legions took on greater pride and identity from the legion they were integrated into than simply being a soldier in a Roman army. The care of the author to
take note of the "1st" called "*Adjutrix*", which was fighting its very first engagement, and the "21st" could very well reveal that the battle expression performed
by these forces in the recorded battle was based on their unit name and number.
For example, massed vocal noise expressing before battle "we are the men of
the 1st the *Adjutrix*!". In response, the 21st could have replied "we are the 21st". The
extract details what each legion did to the other during the battle – the 1st stole
the eagle from the 21st and then the 21st to repair that disgrace slew the legate
and captured many of the standards of the 1st – through recording these details
the author has acknowledged evidence of taunting over the course of the battle
between these two rival legions. The taunting was based on what the legion did
to another Roman legion as a way of professing the fighting qualities of their
military force to boost the morale of their own men as well as to affect the psyche
of their enemy.

Civil war periods, with the devastation and general unwillingness of legions to
fight against other legions, saw the Roman battle expression descend into insulting compatriots and proclaiming the proud origins of legion identity. According to
Tacitus, massed vocal battle expressions, both spontaneous and traditional, were
performed by opposing legions during the AD 69 war between Otho and Vitellius.
Legion and commander glorification, coupled with the denunciation of opposing
legions and their commander, served to successfully generate a volatile and intimidatory atmosphere that stimulated enthusiasm amongst the combatants for battle:

> On both sides was a feeling of shame; on both an ambition for glory. Differ-
> ent exhortations were heard: one side exalted the strength of the legions and
> the army from Germany, while the other praised the high renown of the town
> soldiery and the praetorian cohorts. The Vitellians assailed their opponents
> as lazy and indolent, soldiers corrupted by the circus and the theatre; those
> within the town attacked the Vitellians as foreigners and barbarians. At the
> same time, while they thus lauded or blamed Otho and Vitellius, their mutual
> insults were more productive of enthusiasm than their praise.[170]

In the case of the civil war between Constantine and Maxentius in the early AD 4th
century, the religious symbology adopted by Constantine's forces differed from
that of Maxentius' legions. Constantine's amendments to the military standards[171]
and soldiers' shields,[172] within his army at the battle of Milvian Bridge, permitted
his forces to appear different from their counterparts. According to Lactantius,
Constantine's adoption of the *Chi Rho* symbol reflected a religious invocation to
a God Constantine viewed as the bringer of victory. The willingness of Constantine's troops to alter their traditional shield design suggests several factors. Primarily, the loyalty of the army to Constantine was evident; the troops potentially

viewed the changed appearance of their shields to be beneficial in battle, so as not to be confused with the enemy; the potential support of a powerful deity/spiritual force that was believed to bring victory was viewed as attractive by the troops, who were accustomed to divine invocation and benevolence.

Civil war periods clearly brought with them a host of military problems. One of these was the sterilisation of psychological warfare which aimed to inspire Roman troops and terrify non-Roman enemies. The familiar nature of battle expression meant that traditional military practice would not benefit either army gathered for battle. Significantly, civil war periods demonstrate that the Roman battle expression was culturally unique compared to other military forces of the Graeco-Roman world, whereby common military practices were used across the empire by every legion. Alternatives were sought after and found by way of honouring legion and/or military commander identity, both of which united friendly soldiers against opposing forces. The denunciation of the enemy by way of ethnic/social composition of the legion opposite, inspired from the geographical origin of formation or service, and the belittlement of opposing military commanders, served to overcome the limitations of the battle expression.

Graeco-Roman literary sources reveal that the Romans adopted and manipulated non-Roman military battle expressions to suit their own needs, much like how the Romans adopted and manipulated other cultural aspects from foreign peoples for their own benefit, such as the conversion of Greek deities into Roman equivalents. This is evident in the Roman war cry of Germanic origin, the *barritus*, that imitated the martial custom prevalent among *auxilia palatina* from east of the Rhine.[173] Battle expression of the late empire continued the trend of Roman military forces absorbing non-Roman cultural traits through the incorporation of Christian invocations.

As the empire grew from the early imperial period into the late empire, the integration of non-Roman military units into the army typified the battle expression of the military juggernaut.[174] As seen earlier, civil war periods of the late republic and early empire witnessed non-Roman units within the army adopt traditional customs to differentiate themselves on the battlefield.[175] The progressive barbarisation of the army, especially during the 3rd century is an undeniable fact.[176] Tribal or regional contingents that, during AD 2nd century, had become permanent and almost-regular army units within the army, served far away from their traditional lands in distant corners of the empire, garrisoning their assigned frontier districts for over a century. These non-Roman units may have continued to worship their native gods while serving in the army and were encouraged by the High Command to continue to excel in their native fighting skills. Despite elements of Romanisation in these units, barbarian elite units contributed to the reformation of the traditional Roman army.[177] The integration of non-Roman units represented the nucleus of the nascent field army.[178]

Archaeology reveals that temples, tombstone images and inscriptions testify to the retention of barbarian influence on the Roman army. The Mauri from North Africa served in Dacia and Hadrian's Wall.[179] The enrolment of Sarmatian cavalry into the army for service in Britain,[180] and the incorporation of Germanic warriors within Trajan's army for his Dacian campaign,[181] from the AD 2nd century, reinforces the trend of the imperial period to integrate non-Roman units into the army. Ammianus Marcellinus describes the *Draco*, or dragon-headed military standard, that was adopted by the army from the Sarmatians:

> And behind the manifold others that preceded him he was surrounded by dragons, woven out of purple thread and bound to the golden and jewelled tops of spears, with wide mouths open to the breeze and hence hissing as if roused by anger, and leaving their tails winding in the wind.[182]

The noise generated by the *draco* standard would have been eerie and may help explain why it was integrated into the army. The importance that the Sarmatian *draco* standard held in the army is revealed by Vegetius who claims that those who carried this standard into battle were from a significant officer class, entitled *Draconarii*.[183]

The development of the battle expression, as a result of the integration of non-Roman customs within the army, is elucidated in Ammianus' description of the *barritus*. Due to its effectiveness, in intimidating the enemy and encouraging the participants to fight with greater spirit, became a typical battle expression of the Roman army during the late empire.[184] Modern scholars support this argument claiming the *barritus* imitated the martial custom prevalent among *auxilia palatina* from east of the Rhine.[185] The effect this noise had on the enemy was profound. The overpowering sound and energy that came from the Roman army forced the enemy to flee.[186] A bronze foil from a Germanic helmet unearthed in Denmark may help to demonstrate the nature of the *barritus* cry.[187] From this image, warriors who undertook the *barritus* cry held aloft their shields close to their mouths as they cried aloud. The enhanced sound generated by this contributed to an imposing atmosphere. The reverberations caused by the vocal noise into the shield allowed individuals to determine the strength of their cry, and hence the likelihood of victory.[188]

The Roman army's shift in traditional battle expression types of the early and late empires can be attributed to the problems experienced on the battlefield during periods of civil war. Rome's expansion around the Mediterranean world led to its exposure to non-Roman cultures. Effective elements of non-Roman military units and their battlefield customs, such as the *barritus* and *draco* standard, were integrated into the army. As Rome's socio-political status changed with its territorial expansion, namely the power of the emperor and the rise of the Christian faith, battlefield practices reflected this changing culture. With all these changes, the Roman army was still able to subtly retain traditional battle expression in the process, such as standards: trumpets and massed vocal noise, albeit in consideration of socio-political ideology at the time.

The intentional appearance of ancient military forces is a feature of the battle expression. This military practice was universally adopted by every culture within the confines of the Graeco-Roman Mediterranean world. The appearances forged by armies and warriors often reflected cultural traits of the military force, such as the types of clothing and armour worn, the weapons and equipment carried and hairstyles. The intentional image that ancient armies and warriors presented on the battlefield served to unnerve the enemy, and it attempted to hinder their military potential. For friendly troops, the sight and sound of familiarity aimed to distinguish friend from foe in the heat of battle as well as to instil a sense of confidence and belonging, countering any notion of isolation or weakness in the lead up to and during battle. Evidence suggests that the armies of Sparta, Macedon and Rome had unique identifiers associated with their typical battlefield appearance. These identifiers had links to socio-political and religious traditions that were incorporated into military custom.

Notes

1 Obviously, uniformed appearance held multiple purposes including for identification on the battlefield to distinguish friend from foe.
2 Sabin (2007): 421. In reference to the appearance and function of exotic weapons, Sabin stresses the psychological impact that terrifying sights, smells and sounds had in spreading terror within an unprepared enemy army on the battlefield.
3 Hd. 7.69.
4 Xen. *Const. Lae.* 11.3.
5 For opponents of Spartan armies being reluctant to fight, see Xen. *Hell.* 4.2.18; 4.4.16, 4.6.11; Plut. *Lyc.* 22.2–3.
6 Xen. *Const. Lae.* 13.8–9.
7 Plut. *Alex.* 16.4; Arr. *Ana.* 1.14.4.
8 Arr. *Ana.* 3.13.1–2.
9 Arr. *Ana.* 1.14.4. To see dread on enemy faces at the sight of Alexander leading his army into the thick of battle, see the Alexander Mosaic from the House of the Faun Pompeii.
10 Polyb. 6.23.
11 Veg. *DRM.* 2.14; Maur. *Strat.* 7.2.15.
12 Veg. *DRM.* 2.16.
13 As was the case for the Marius led Romans against the Teutons and Ambrones, see Plut. *Mar.* 15–16.
14 Veg. *DRM.* 3.12.
15 Livy. 38.17.5.
16 Livy. 38.17.3–5. See also Plut. Mar. 15–16.
17 Livy. 23.16.14–15.
18 Plut. *Mar.* 15–16.
19 Plut. *Mar.* 15–16.
20 Plut. *Mar.* 19.
21 Plut. *Mar.* 19.
22 For example, the Greeks with their Paean and the Romans with their trumpets, Plut. *Lyc.* 22.2–3; Sal. *Jug.* 99.
23 Speidel (2004).
24 Kinsella, Thomas, trans. *"The Táin"* Oxford: Oxford University Press, 1969: 150–153.
25 Ulster Journal of Archaeology. "War-Cries of Irish Septs" Ulster Journal of Archaeology, First Series, Vol. 3 (1855), pp. 203–206.
26 Speidel (2002): 269, 272.
27 Ulster Journal of Archaeology. "War-Cries of Irish Septs" Ulster Journal of Archaeology, First Series, Vol. 3 (1855), p. 206.
28 Amm. 19.2.11.

29 Amm. 19.2.11–12.
30 Plut. *Lyc*. 21.1–4.
31 Plut. *Lyc*. 22.2–3.
32 For more detail on the impact Tyrtaeus had on Spartan military custom and society, see Bayliss (2017).
33 Plut. *Inst. Lac*. 16; Bayliss (2017): 63.
34 Ath. *Dei*. 14.630.
35 Ath. *Dei*. 14.630.
36 See Bayliss (2017); Banks (1853): 327–343.
37 Tyr. 3.
38 Moosavi (2015) provides a good outline of literature types produced during wartime that sponsored the concept of martyrdom in a military context for the benefit of religion, nation and culture.
39 Moosavi (2015): 9.
40 Moosavi (2015): 9.
41 Moosavi (2015): 10.
42 Moosavi (2015): 10.
43 Moosavi (2015): 11–13.
44 Tyr. 1.
45 Tyr. 1–3.
46 Tyr. 6. The pious nature of Spartan military practice, such as the singing of the paean, customary sacrifices of the she-goat and the playing of the hymn to Castor on the flute, helped the Spartan warriors believe that heaven was on their side. See Plut. *Lyc*. 22.3.
47 Tyr. 2.
48 Bayliss (2017); Cartledge (2006): 79; Kõiv (2005): 238, 263; Ath. *Dei*. 14.630ff; Plut. *Cleo*. 2; Plut. *Mor*. 959a.
49 Bayliss (2017): 79.
50 Bayliss (2017): 63–65.
51 Thuc. 5.69.
52 Xen. *Const. Lac*. 11.3.
53 Plut. *Inst*. 24. The red cloak served to camouflage wounds sustained on the battlefield.
54 Xen. *Const. Lac*. 13.8–9.
55 Xen. *Const. Lac*. 12.5 & 13.8–9.
56 Xen. *Const. Lac*. 12.5 & 13.8–9.
57 Xen. *Const. Lac*. 12.5 & 13.8–9.
58 Thuc. 5.70.
59 Plut. *Lyc*. 22.2.
60 Plut. *Lyc*. 22.3.
61 Hdt. 9.61–62.
62 Plut. *Ages*. 36.
63 Spier (1990): 114.
64 Spier (1990): 127.
65 Rusch (2011): 207.
66 Aesch. *Sept*. 87–90, 380–399, 425–434, 486–498; Polyb. 4.64.6.
67 Xen. *Hell*. 7.5.20. This reference details Arcadian warriors duplicating the Theban club on their shields before battle.
68 Eupolis. 394. "Seeing the flashing Lambdas he was terrified".
69 Rusch (2011): 31.
70 Xen. *Hell*. 4.4.10.
71 Dio. Sic. 16.3.1–3.
72 Arr. *Ana*. 7.12.2.
73 Arr. *Ana*. 1.6.1–4.
74 Arr. *Ana*. 1.14.7; Xen. *Ana*. 1.8.18.

75 Poly. *Strat. Alex.* 4.3.5.
76 Foss (1975): 27–28.
77 Foss (1975): 28.
78 Kelly (2012): 282.
79 Roman military triumphs generally commemorated single military engagements, such as a siege that lasted many months or a battle that lasted a few hours. The main criteria involved a victory that resulted in the deaths of five thousand of the enemy (Val. Max. 2.8.1). A triumph could commemorate a military campaign, with particular focus on one or a number of key battles. For example, the triumph of Titus and Vespasian after the Jewish War in AD first century – which has been immortalised with the Arch of Titus in Roman Forum complex specifically focused on the AD 70 siege of Jerusalem, however, was also used to acknowledge the role of Vespasian, who led the campaign during Nero's reign, however, after the death of Nero and the rise of Vespasian as emperor, the triumph Titus and Vespasian undertook embraced a number of social, political and military factors that aimed to solidify the Flavian rule over the Roman world.
80 Beard (2007): 306; Eder (2006); Varro *DLL*. 6.68.
    See also Warren (1970): 49–66; Erskine (2013): 37–55 and Armstrong (2013): 7–21 for a more detailed overview of the origins of the Roman triumph.
81 Beard (2007): 314–318.
82 D. H. *Ant.* 2.34.
83 For more on battlefield oaths, see Chapter 8.
84 Plut. *Marc.* 8. Livy 4.20 also refers to Jupiter Feretrius being worshipped during a military triumph.
85 Plut. *Sul.* 19.
86 See Brilliant (1999): 225; Warren (1970): 65; Armstrong (2013): 11; Eder (2006).
87 Beard (2007): 243. For examples of commanders being hailed "imperator", see; Polyb. 10.40; Livy. 45.38.12; Plut. *Cras.* 17.2; Plut. *Pom.* 12; Jos. *BJ.* 6.316.
88 Varro. *DLL*. 6.68.
89 Beard (2007): 244–245. See also Armstrong (2013): 11–13.
90 Livy. 45.38.12 & Beard (2007): 246.
91 Livy. 4.53.11–13.
92 Livy. 4.20.2; Beard (2007): 247.
93 Beard (2007): 244–245.
94 Beard (2007): 248–249. See n.86 p. 249.
95 Livy. 5.49.7.
96 Livy. 3.29.5.
97 Suet. *Jul.* 49.4.
98 Suet. *Jul.* 49.4.
99 Suet. *Jul.* 51.
100 For examples of studies, see Beard (2007); Warren (1970); Brilliant (1999); Armstrong (2013); Erskine (2013); Eder (2006).
101 Tac. *Hist.* 2.43; 3.9; D. H. *Ant.* 11.43; Jos. *BJ.* 3.123; Tac. *Ann.* 2.17; Suet. *Cal.* 14; Caes. *Gal.* 4.25.
102 D. H. *Ant.* 6.45, 10.16. Jos. *BJ.* 6.316 also refers to the religious significance of the Roman standards.
103 D. H. *Ant.* 9.31, 9.50, 11.43.
104 Tac. *Hist.* 3.9.
105 Jos. *BJ.* 6.403.
106 Jos. *BJ.* 6.316.
107 Speidel (1984): 17–22.
108 Speidel (1984): 17–18.
109 Jos. *BJ.* 3.123; Tac. *Ann.* 2.17.

110 D. H. *Ant.* 10.36 for an example of the respect afforded to those that carried the eagle. See also, OCD 3rd ed. (1997): 1395–1396.
111 Speidel (1984): 22.
112 Suet. *Claud.* 13; Speidel (1984): 17, 20–22.
113 Jos. *BJ.* 3.123; Tac. *Ann.* 2.17; Speidel (1984): 20–21.
114 Amm. 16.12.12–13.
115 Amm. 27.10.9.
116 Amm. 27.10.9.
117 Livy 28.14.10.
118 Amm. 27.2.6.
119 Amm. 28.5.3. See also Amm. 29.5.15.
120 For the different variety of trumpets and bronze instruments, see Veg. *DRM.* 2.7 & Gleason (2008): 231–232.
121 For examples of Roman trumpet use for issuing orders not related to battle expression, see Plut. *Pomp.* 70; Plut. *Sul.* 29; App. *BC.* 5.4.38; Livy. 27.15.14; Caes. *B. Civ.* 3.46; Amm. 20.11.8; Jos. *BJ.* 3.86–91.
122 The literary sources do not detail the lyrics or noise associated with these massed vocal sounds initiated by trumpet but are often translated as shouts or cries. Given the limitations of the term war cry/battle cry, these vocal noises could very well be song.
123 Livy. 10.40.14, 23.16; Livy. 25.37, 25.39.3, 30.33.13; Livy. 33.9.1–2.
124 Livy. 23.16.12.
125 D. H. *Ant.* 8.84. See also: D. H. *Ant.* 6.10.2 & 9.11.1.
126 Caes. *B. Civ.* 3.90.
127 Sal. *Jug.* 99. "Marius . . . ordered the watchmen and likewise the horn blowers of the cohorts, of the cavalry squadrons and of the legions to sound simultaneously, and without warning, all their signals, and the soldiers to raise a shout".
128 Jos. *BJ.* 3.265.
129 Plut. *Sul.* 14.3. "Blasts of many trumpets and bugles, and by the cries and yells of the soldiery".
    See also: Plut. *M. Cato.* 13.7. "with bray of trumpet and battle-cry".
130 Amm. 21.12.5. "The sound of the trumpets roused them to slay one another, and raising a shout they rushed to battle". See also: Amm. 24.4.15 "the trumpets sounded their martial note, both sides raised a loud shout" & Amm. 31.13.1. "On every side armour and weapons clashed, and Bellona, raging with more than usual madness for the destruction of the Romans, blew her lamentable war-trumpets; our soldiers who were giving way rallied, exchanging many encouraging shouts".
131 Polyb. 15.12.8. Also referred to in Livy 30.33.13.
132 For examples, see Livy. 25.37.10–12; 33.9.1–2; Plut. *Sul.* 14.3; Sal. *Jug.* 99.
133 Sal. *Jug.* 99.
134 Speidel (1984): 33.
135 App. *BC.* 2.11.78.
136 Amm. 24.5.9.
137 Amm. 27.10.12.
138 Cowan (2007); Rance (2015).
139 Rance (2015): 1.
140 Dio. Cass. 49.9; Veg. *DRM.* 3.18; Amm. 16.12.43; 21.13.15; Procop. *Wars.* 4.11.36.
141 Veg. *DRM.* 3.18.
142 Cowan (2007): 114–115; see also Chapter 2.
143 Caes. *B. Civ.* 3.92.
144 D. H. *Ant.* 8.84.
145 D. H. *Ant.* 8.66, 9.70, 10.21, 10.46.5.
146 D. H. *Ant.* 9.11.
147 Plut. *Marc.* 8.

148 D. H. *Ant.* 8.66.
149 Had the battle expression not been successful or lacked enthusiasm, surely the efforts of the men to continue to attempt it would have faded out.
150 D. H. *Ant.* 9.70.
151 Ono. *Strat.* 29.
152 Ono. *Strat.* Pr.8.
153 Amm 19.2.11.
154 Such as Whately (2016); Cowan (2007); Rance (2015); Speidel (2004).
155 Along with other typically "English" tunes sung by travelling supporters of English sporting teams such as the English cricket and football teams.
156 This practice should also be applied to the Persian social tradition too. The Roman practice of clients addressing patrons as *Dominus* supports this idea.
157 Or Sapor.
158 Roman slingers were integrated within the army no later than the middle of the 2nd century BC.
159 Weiß (2006).
160 Sling bullets, naturally, focused on physically harming the enemy. The battle expression aimed to destroy the confidence levels of the enemy. This psychological advantage over the enemy sought after hoped to affect the military capabilities of the enemy with the purpose of influencing their retreat from the battlefield before the advance to engage, or with the hope of generating an atmosphere where the enemy would be destroyed upon violent engagement.
161 Weiß (2006) Octavia(ni) culum peto *"I am aiming at Octavian's backside"*.
     Fulviae [la]ndicam peto *"I am aiming at Fulvia's clitoris"*.
     For more information regarding sling bullets unearthed from the Perusine War, see Hallett (1977).
162 By way of language used in a military context and intent to inflict physical damage on the enemy using disrespectful, derogatory sentiments. See Hallett (1977): 151–154.
163 Specifically, Antony's brother and his wife Fulvia on his behalf.
164 Hallett (1977): 154.
165 Hallett (1977): 155.
166 App. *BC.* 3.9.68.
167 Tac. *Hist.* 3.22.
168 Tac. *Hist.* 5.16–17.
169 Tac. *Hist.* 2.43.
170 Tac. *Hist* 2.21.
171 Eus. *VC.* 1.28–31; Bruun (1997): 41–42.
172 Lac. *DM.* 44.5.
173 Rance (2015): 1.
174 For a comprehensive examination of the Roman *auxilia* between the reigns of Augustus and Alexander Severus, see Haynes (2013).
175 This is particularly the case with the Batavian Rebellion of 69–70 and the reference; Tac. *Hist.* 5.16–17.
176 Speidel (1984): 118. For research surrounding recruitment of the Roman army, see Dobson & Mann (1973).
177 Speidel (1984): 118–123.
178 Speidel (1984): 128.
179 Speidel (1984): 124–125.
180 Dio. Cass. 72.16.1–2. See also: Littleton & Thomas (1978): 520; Wadge (1987): 209.
181 For a study on the Germanic warriors depicted on Trajan's column, see Speidel (2004).
182 Amm 16.10.7.
183 Veg. *DRM.* 2.7.

184 Amm. 26.7.17. See also Amm. 16.12.43; 21.13.15; 31.7.11; Speidel (2004): 110–111.
185 Rance (2015): 1.
186 Amm. 21.13.15.
187 Speidel (2004): 112. Figure 10.1 (Helmet from grave VII, Valsgärde, Uppland, Denmark).
188 Speidel (2004): 111.

8   Battlefield oaths

Military oaths recited by individuals, military units or entire armies on the battlefield is another dimension of the ancient Graeco-Roman battle expression. Oaths were covenants made between two or more parties that sought to generate benevolence for each party associated. In the military context, battlefield oaths served to ensure victory and preserve the honour for the oath taker/s while guaranteeing the fulfilment of a material promise made to the party (often a deity) that facilitated the oath taker/s victorious endeavours. Graeco-Roman literary evidence reveals that Greek battlefield oaths were commonly socio-religious in nature. Oath takers publicly made a pact with patron deities that they be a witness to the pledges made and that actions promised will be fulfilled, such as defeating the enemy unto death or to continue fighting until death. Roman battlefield oaths generally contrasted with their Greek counterparts. Roman battlefield oaths aimed to invoke the benevolence of targeted deities to guarantee victory for the oath takers. In recompense for victorious support, oath takers would pledge to glorify the deity in question with the dedication of a religious sanctuary (temple, shrine) or initiate commemorative ritual for the worship of the subjected deity.

The Spartans are claimed to have recited the oath of the "sworn bands" from the late 6th century BC onwards.[1] There are clear links between this oath and the so-called "Oath at Plataea" that was said to have been sworn at the Isthmus prior to the Battle of Plataea in 479 BC.[2] The Spartan sworn band was a squad of 40 soldiers bound together by a solemn oath. There are clear elements of the Oath of Plataea that originated from the oath of the sworn bands.[3] These include the promise to fight to the death for the freedom of Sparta, a pledge of loyalty to the officer class and fellow soldiers in life and death and a promise to provide dead comrades with proper burial rites. Each of these elements was rooted in the writings of Tyrtaeus.[4]

The oath sworn by the "Seven" in Aeschylus' *Seven Against Thebes* generally complies with the understanding of ancient Greek battlefield oaths. That is deities were used as witnesses to a pledge made. The oath itself is not based on acquiring the deities' support in achieving victory, but rather using them as witnesses to the promise that victory will be achieved, or the seven oath takers would die attempting to achieve the pledge:

DOI: 10.4324/9781003280439-8

Seven men, bold leaders of companies, slaughtered a bull, let its blood run
into a black-rimmed shield, and touching the bull's blood with their hands
swore an oath by Ares, Enyo, and blood-loving Terror, that they would either
bring destruction on the city, sacking the town of the Cadmeans by force, or
perish and mix their blood into the soil of this land; and with their own hands,
shedding tears, they were adorning the chariot of Adrastus with mementoes
of themselves to take home to their parents.[5]

This type of oath recorded in Aeschylus' play is both recognisable yet uncanny in
nature.[6] The oath takers slaughtered a bull over a black-rimmed shield, touching
the animal's blood with their hands. They invoke Ares and Enyo and blood-loving
Terror and swear that they will destroy the city and sack the town of the Cadmeans
by force or die making the earth a paste with their blood. They send mementoes
home to their parents, presumably locks of their hair shedding tears in the process.
The nature of this battlefield oath is unprecedented. Many features of the Seven's
oath are recognisable in non-military-related oath taking such as law-court oath processes and male initiation into adulthood (ephebic oath). However, for oath takers to
invoke a triad of exclusively militaristic deities – Ares, Enyo and Phobos (Terror) –
and the physical contact with the oath victim's blood is unique.[7] It is not surprising
that these militaristic gods are associated with the battlefield oath of the Seven, as
these gods are invoked in other types of Greek battle expression, such as the paean.

Roman generals and armies prior to battle made oaths and dedications to their
deities in exchange for victory. Public oaths dedicated to the gods galvanised the
army to fulfil covenants made. Graeco-Roman literary sources suggest there was
a process in the creation of a military oath. An oath was usually initiated by the
commander of the army, who would invoke a god – whether the god selected
was designated due to the time of year; the situation at hand; the commander's
knowledge of deities and their roles, remains to be seen. The pledge made by the
commander to the god would be pronounced publicly: the dedication of a temple
structure; the development of a priestly order; a promise to be loyal/to defeat the
enemy/to never surrender; the dedication of booty, was then made and supported,
almost in a choreographed sequence, by the army via some kind of united massed
vocal acclamation in support of the selection of god and covenant made. Onosander's *Strategikos* dedicates a chapter to "Propitiation of the divine power by
the general before leading the army into battle". This military handbook clearly
instructed Roman generals that religious obligation to the gods based on what the
law dictated was a necessary requirement for gaining the influence of the gods for
any military enterprise.[8]

The piety shown by Roman military forces through this form of battle expression reveals a level of spiritual meaning and sophistication by way of the selection process of deity and dedication promised. One such example is connected to
Brutus' oath made to overthrow and defeat the tyrannical reign of the Tarquinii:

He swore by Mars and all the other gods that he would do everything in his
power to overthrow the dominion of the Tarquinii and that he would neither

be reconciled to the tyrants himself nor tolerate any who should be reconciled to them, but would look upon every man who thought otherwise as an enemy and till his death would pursue with unrelenting hatred both the tyranny and its abettors; and if he should violate his oath, he prayed that he and his children might meet with the same end as Lucretia.

Having said this, he called upon all the rest also to take the same oath; and they, no longer hesitating, rose up, and receiving the dagger from one another, swore.[9]

Brutus' oath, despite not being on a field of battle, but in the house where Lucretia had committed suicide, marks the prelude to the uprising against the last king of Rome. Its military context is evident, and this oath may have served as inspiration for later Roman armies when confronted with similar tyrannical military opponents.

During the early republic, the typical military oath was to stand by and remain loyal to the standards and the commander.[10] The nature of oath taking involved the oath maker holding aloft their sword, or primary weapon, and swearing to specific gods by name and detailing the covenant that would be made between them. One example of such a process is of the oath taken by Flavoleius:

he held up his sword and took the oath traditional among the Romans and regarded by them as the mightiest of all, swearing by his own good faith that he would return to Rome victorious over the enemy, or not at all.[11]

What is interesting in this extract is that the oath made by Flavoleius is referred to as traditional among the Romans (ἐπιχώριόν τε Ῥωμαίοις), suggesting that this oath was commonly used in military contexts during the early republic. There were other types of oaths made in military contexts by the Romans. The Flavoleius oath, to return to Rome victorious or not at all, is dubbed as "the mightiest of all" (κράτιστον ὅρκον), suggesting there were a variety of military oaths that could be made before the onset of battle. Of note, in this reference, the oath maker is not the commander-in-chief of the army, but rather a *primipilus*, or senior centurion of a legion.[12] Flavoleius was a well-respected man within the rank and file of the military and after he had made this vow inspired the rest of the military force to do likewise which caused great inspiration to fill the Roman army:

After Flavoleius had taken this oath there was great applause from all; and immediately both the consuls did the same, as did also the subordinate officers, both tribunes and centurions, and last of all the rank and file. When this had been done, great cheerfulness came upon them all and great affection for one another and also confidence and ardour.[13]

Following the oath inspired by Flavoleius, the consuls, who were in command of the army, further invoked the support of the gods by "vows, sacrifices, and prayers to be their guides as they marched out, led the army out of the camp in regular order and formation".[14]

The order of the *Salii* was developed during the reign of King Hostilius in the 7th century BC in accordance with a vow made in a war against the Sabines. The role of the *Salii* was to dance and sing to hymns in praise of the gods of war.[15] From this narrative, it is understood that an oath or invocation of war gods was made, presumably in the lead up to the battle between the Romans (led by King Hostilius) and the Sabines. The covenant made between King Hostilius and the war gods of Rome was the extension of the *Salii* order, in which a key feature was for the *Salii* to dance and sing hymns in honour of the war gods in commemoration of their military victory over the Sabines.

The promise to construct a temple dedicated to a specific deity was another form of the military oath made by Roman commanders. Aside from the public building to be constructed, these types of oaths provide information on which gods were viewed as Roman war gods. Livy details an oath made by consul C. Cornelius Cethegus in ca. 197 BC in the lead up to battle against the Insubres. In this oath, the consul pledged:

> a temple to Juno Sospita if the enemy should be routed and put to flight that day; the soldiers raised a shout affirming that they would ensure the fulfillment of the consul's prayer, and the attack on the enemy began.[16]

The promise to erect a temple dedicated to the goddess Juno, who, with the epithet "*Sospita*" was deemed to be a "saviour" within a military context. In this capacity, Juno *Sospita* was a war goddess. The Roman army, by raising a shout that affirmed the consul's pledge suggests that they were supportive of the oath made and of the deity invoked.

Another form of Roman military oath was the practice of promising to offer up an enemy suit of armour to the gods as a means of gaining their benefaction. According to Plutarch, before the battle against the Gauls, led by their king Viridomarus, Marcellus vowed that he would consecrate to Jupiter Feretrius the most beautiful suit of armour among the Gallic host.[17] Plutarch explains that Jupiter may have been given the epithet, *Feretrius*, due to the Latin word *ferire*, which means to smite – in reference to Jupiter's thunder-wielding capabilities. The epithet, according to Plutarch, may also originate from the Roman exhortation to each other on the battlefield when in pursuit of a fleeing enemy to *feri* or "strike".[18] During the ensuing battle narrative, Plutarch claims that the Gallic king (Viridomarus) challenged Marcellus to single combat in which Marcellus determined that the armour worn by this king was the finest within the enemy army and accepted combat. After defeating the enemy king, Marcellus leapt from his horse:

> Laying his hands upon the armour of the dead, he looked towards heaven and said: "O Jupiter Feretrius, who beholdest the great deeds and exploits of generals and commanders in wars and fightings, I call thee to witness that I have overpowered and slain this man with my own hand, being the third Roman ruler and general so to slay a ruler and king, and that I dedicate to thee the first

and most beautiful of the spoils. Do thou therefore grant us a like fortune as we prosecute the rest of the war.[19]

Of note, Plutarch details the fulfilment of the oath of Marcellus whereby during the triumph held in Rome to commemorate the victory in battle over the Gauls, the suit of armour was displayed and deposited in the temple of Jupiter Feretrius.[20] Through the personification of Jupiter and his epithet Feretrius, Roman commanders were able to discern amicable Roman deities against foreign enemies on the battlefield to use for inspiration and benefaction.[21]

Further evidence to support the notion that Romans in the lead up to battle sought the positive intervention from the gods by way of oath taking is revealed through Plutarch's *Life of Sulla*.[22] According to Plutarch:

> Sulla had a little golden image of Apollo from Delphi which he always carried in his bosom when he was in battle, but that on this occasion he took it out and kissed it affectionately, saying: "O Pythian Apollo, now that thou hast in so many struggles raised the fortunate Cornelius Sulla to glory and greatness, can it be that thou hast brought him to the gates of his native city only to cast him down there, to perish most shamefully with his fellow-countrymen?" Thus, invoking the god, they say, he entreated some of his men, threatened others, and laid hands on others still; but at last his left wing was completely shattered.[23]

In this extract, rather than Sulla making a covenant with Apollo for victory, Sulla attempted to, in public view of his soldiers, invoke the god – or at least be seen to be invoking the god – by reminding those within earshot of his past military success, due to the aid of the deity in question. Sulla aimed to strengthen the fighting determination of his distressed left wing by appearing to invoke Pythian Apollo. Unfortunately, for Sulla, his attempt to stir his men into successful action did not work. Indeed, what is clear from this extract is that Roman soldiers could carry on their personal amulets or statuettes of deities into battle. Furthermore, Roman soldiers could use a personal deity as a means for inspiration in the lead up to battle, as Sulla did. It could also be asserted that individual soldiers had the ability to create their own covenants with select divinities and that those soldiers within their immediate section of the battle line may bear witness to these oaths.

The practice of Roman soldiers swearing an oath to Roman gods, as a battle expression, continued into the late empire. Ammianus Marcellinus details the period of hostility between Procopius and Valens.[24] On the battlefield, in the lead up to an engagement between two Roman military forces, Ammianus claims that Procopius (leading one of the Roman armies) ventured out into the middle of the battlefield as if to challenge the enemy. However, instead of single combat, Procopius made a gesture of peace with the opposing commander, Vitalianus.[25] According to Ammianus, Procopius was so successful in preventing the two Roman armies from fighting each other that they, instead, joined forces under the command of Procopius.[26] It was once the armies had joined forces that Ammianus details an example of a military oath. In the celebratory atmosphere on this would-be battlefield, Ammianus states that the soldiers:

> Swearing in the soldiers' manner by Jupiter that Procopius would be invincible.[27]

The battle expression detailed in this extract relates to the Roman soldiers swearing an oath to Jupiter that their designated commander would be victorious in his upcoming war. Even though this episode did not result in a battle, what is significant, is that Ammianus claims that this oath was customary to Roman soldiers in a battle context; *testati more militiae*.

The late empire saw great alterations to the Roman military, by way of the integration of non-Roman units and customs into the army. For example, in the verses that precede the extract above, Ammianus details the arguments put forward by Procopius which convinced the armies not to fight each other. These arguments stressed the unity of Rome and war against foreigners.[28] Ammianus details that instead of the *barritus*, which was an adopted Germanic battle expression reserved for the heat of battle, the soldiers adopted the customary Roman battle expression of taking an oath to Jupiter. As a result of Procopius' arguments, the Roman armies were restored along ethnic cultural lines, which culminated in the undertaking of a typical battle expression, the military oath.

AD 5th-century author, Vegetius, demonstrates that the tradition of military oaths continued into the Christian era. In his work, *De Re Militari*, Vegetius records the military oath legionaries took once they enrolled into the army and received the military mark on their hands:

> They swear by God, by Christ and by the Holy Ghost; and by the Majesty of the Emperor who, after God, should be the chief object of the love and veneration of mankind. . . . The soldiers, therefore, swear they will obey the Emperor willingly and implicitly in all his commands, that they will never desert and will always be ready to sacrifice their lives for the Roman Empire.[29]

The similarities between the late empire and the republican military oath are clear, revealing a tradition that did not vary greatly. The promise to remain loyal to the commander of the military and to never desert the standards is a clear commonality. The main difference is the political and religious orientation of Rome during the republic and late empire. The political rule of the emperor in the late empire compared to the senatorial system of the republic coupled with the rise of Christianity over the pantheon of Roman war gods of pre-Christian times testify to the change that had infiltrated the Roman army but did not alter the oath tradition.

From the Graeco-Roman literary record, oaths made by Roman armies were a characteristic feature of military life in a battlefield context that continued

---

from the early republic through to the late empire. Publicly making oaths in the lead up to battle demonstrates the close connection Roman religion had to military practice. The unique process that Roman commanders and armies recited oaths separates them from other cultural groups referred to in Graeco-Roman literary sources. The sophistication evident in the creation and recital of military oaths reveals religious observance was important to the army. Also, the high command's ability to interpret a military situation, by selecting which deity would be invoked and what covenant would be made in coordination, through acclamation, with the rank and file of the army denotes sophistication.

Greek and Roman armies were familiar with battlefield oaths. Literary evidence suggests that oaths were employed in a military context to legitimise guarantees made in relation to victory and loyalty to the fighting group. The Greeks and the Romans generally differed in the use of battlefield oaths. Greek oaths invoked the gods to bear witness to ensure the agreements made by word or sacrifice would be honoured with only death breaking the pledge. Roman oaths, on the other hand, sought to entice the support of the gods to achieve support military success for the specific god/s to gain greater public acclaim through the dedication of honorific structure or custom.

Notes

1 Van Wees (2006): 135.
2 See Diodorus. 11.29.2–4; Lycurgus, *Against Leocrates*, 81; Van Wees (2006). For more on Greek oaths and the Oath of Plataea, see the works of Krentz (2007) and Sommerstein & Bayliss (2017).
3 Van Wees (2006): 135.
4 Van Wees (2006): 151–153.
5 Aesch. *Sept.* 42–51.
6 Torrance (2015): 282.
7 Torrance (2015): 282.
8 Ono. *Strat.* 5.
9 D.H. *Ant.* 4.70–71.
10 D. H. *Ant.* 6.45; Livy. 22.38.2–3.
11 D. H. *Ant.* 9.10.4.
12 D. H. *Ant.* 9.10.2.
13 D. H. *Ant.* 9.10.4–5.
14 D. H. *Ant.* 9.10.6.
15 D. H. *Ant.* 2.70.
16 Livy. 32.30.10.
17 Plut. *Marc.* 6–8.
18 Plut. *Marc.* 8.
19 Plut. *Marc.* 7.
20 Plut. *Marc.* 8.
21 Despite Polybius not referring to this episode in his earlier recording of this battle – which may suggest Plutarch's embellishment of the exact circumstances – it is still a useful source when focusing on the use of oaths, the deities' Roman military forces sought support from, and the types of covenants made on the battlefield as a Roman battle expression.

22 Plut. *Sul*. 29.
23 Plut. *Sul*. 29.
24 Amm. 26.7.15–17.
25 Amm. 26.7.15–16.
26 Amm. 26.7.17.
27 Amm. 26.7.17.
28 Amm. 26.7.16.
29 Veg. *DRM*. 2.5.

9   Conclusion

The sophistication and cultural meaning exhibited in planned and spontaneous battle expression forms expose the limitations of the term war cry. Only by integrating the information contained within Graeco-Roman literary source material and supporting archaeological remains can we begin to account appropriately and effectively for a far broader ancient military phenomenon. To fully comprehend the purpose, significance, range and typology of battlefield customs available to armies of the Graeco-Roman Mediterranean world, a fresh approach needs to be adopted. The term battle expression attempts to re-establish an ancient military phenomenon that has been misunderstood and only partly acknowledged by modern scholarship, and largely misrepresented in modern-day media forms.

Graeco-Roman authors and ancient artworks produced before the 5th century BC through to AD 6th century provide reliable accounts for the battle expression despite the authors' subjective perceptions concerning the cultural uniqueness of battle expression types. African battle expression forms were associated with peculiar action and reflected the multilingual dynamics of that geographical region. Asian forms were presented as ostentatious affairs which exhibited flair and flamboyance pertaining to size, movement and colour. Celtic/Germanic battle expression types were typically linked to their animistic cultural origins. Greek forms portrayed pious endeavours and proclamations of socio-political ideologies reminiscent of the poleis structure. Roman battle expression types varied over the course of their influence around the Mediterranean world. Traditional practices, established during Rome's early foundation, developed and diversified in collaboration with their imperial expansion. During the late empire, Roman battle expression types embraced non-Roman practice and sentiment reflective of the ethnic composition of the army.

There were homogenous features of the battle expression found amongst the different cultural groups of the Graeco-Roman world. The practice of undertaking battlefield customs was universal across cultural groups. The attempt to unite an army through establishing confidence and group cohesion using familiar action was a common feature within all battle expression types. The cohesive manner in which battle expression types were performed underpins the collective aim to intimidate and adversely affect the enemy, a consistent feature found amongst all cultural groups. Spontaneous taunting of the enemy to alter the psyche of those

DOI: 10.4324/9781003280439-9

on the battlefield, the intentional attempts to appear distinctive to the enemy, and the military high commands' encouragement to undertake forms of battle expression on the battlefield highlight universal traits found within the battle expression.

It is undeniable that armies from the ancient Graeco-Roman Mediterranean world expressed themselves *en masse* through a variety of forms on the battlefield. Armies aimed to intimidate their enemy and enthuse their own force through inspiring and terrifying vocal, musical and/or gesticular displays. The culturally unique methods that armies adopted to achieve these displays reflected aspects of their society in a military context. Careful study of the battle expression reveals important information about various cultural groups. The battlefield terrain was intentionally or coincidentally exploited by ancient military forces to generate ideal atmospheric conditions for military objectives. Literary evidence reveals that the impact a battle expression had on an enemy or participating troops could be influenced by the atmosphere created. This battlefield atmosphere could be artificially manufactured through massed noise, silence or movement that targeted the human senses. Modern-day comparisons of this type of atmosphere may be drawn from the choreographed and spontaneous performances of European[1] football supporters inside stadiums.

Each cultural group within the ancient Graeco-Roman world expressed aspects of their spirituality on the battlefield. A common feature amongst each military force was the attempt to invoke the benevolence and intercession of their military-oriented deities on the outcome of the battle. Jugurtha forces in North Africa undertook peculiar gesticulations during battle to communicate with distant allies. These movements appear to express significant cultural practice relating to faith in a military context. The intentional appearance of armies on the battlefield similarly captures culturally specific aspects of spiritual belief. Ethiopian warriors adorned themselves in animal skins and painted their bodies. Spartan hoplites wore garlands in their hair and sacrificed a she-goat in full view of the enemy. The Macedonian army knelt in prayer before the battle of Issus. Celtic/Germanic warriors associated themselves with specific warrior styles that held spiritual connections to natural forces and beings. The battlefield customs of military forces drawn from the peoples of Asia reflected their religious diversity. Trojan armies invoked their deities through cries, likened to a flock of migrating birds. Displaying the severed heads of slain enemies to the enemy and friendly forces during the battle, in accompaniment with ritualised dancing, was a prominent feature of cultural groups from northern Asia Minor and elsewhere. The singing of religious hymns (the *paean*) and blood sacrifice were typical of Greek pre-battle custom. The continuance of archaic religious practice, by Roman forces, reveals significant understandings of military-associated deities and the close affiliation religious tradition played within the military. The adoption of Christian battle expression types during the late empire highlights the long-standing Roman military tradition of employing religiously inspired customs in battlefield contexts.

The acclamation of socio-political ideology on the battlefield, by way of praising and acknowledging political/social superiors and recounting fundamental doctrine, were typical features of the battle expression of cultural groups. Close examination of literary and archaeological source material reveals that Roman, Greek, Celtic/Germanic and Asian cultures took inspiration from expressing their socio-political orientation on the battlefield. By honouring the eagle and other military insignia, Roman forces encapsulated their pride in their socio-political origins. The acknowledgement and praise (or in some cases disrespect in the context of a military triumph) of military commanders reveal key teachings regarding the socio-political features of Rome. Similarly, Asian armies, notably Persian and Parthian, proclaimed the title of their king as *king of kings* (*Shahanshah*), reflective of customary protocols and the imperialistic orientation of their society. The Macedonian army inscribed the name of their king on lead sling bullets. Whether this was done vocally en masse by the army at large during battle remains to be seen; however, it seems more than likely given the time and effort taken in producing missile weapons adorned with socio-political ideology. The significance of identifying with a cultural group and the inspiration taken from it is highlighted through the inscriptions found on lead sling bullets. The battle expression aimed to strike the enemy psychologically and, in the case of lead sling bullets, physically. So, too, we see the battle expression articulating an army's purpose for fighting as an instantiation of its socio-political ideology. *Polis* armies of Greece identified their socio-political origins through the singing of *paean* hymns and shield motifs. The Spartan army recited poetry and song on the battlefield that emphasised foundational aspects of their militaristic society. Celtic/Germanic tribal armies proclaimed kinship groups and ancestral origins on the battlefield through unique battle expression types.

Armies from the Graeco-Roman world utilised battle expression types as an added dimension to their overall military strategy. The intimidatory and provocative nature of battle expression types aimed to influence the course and, as a result, the outcome of a battle before the violent confrontation could occur. Ancient armies adopted vocal and instrumental sound as well as movement and appearance to strike fear and terror into the enemy, with the aim of weakening their fighting capacity. Battle expression types that proved successful, at distorting the mindset and strategy of the enemy, could influence whether the enemy fled from the battlefield, resulting in a bloodless victory, or whether the enemy would abandon their preconceived military strategy, leading to unintended action. The encouragement from the military high command to undertake battle expression reveals its multidimensional purpose. The performance of a battle expression could influence the tactics adopted by the military high command. A weak battle expression by way of cohesion and noise generated in comparison to a more effective battle expression from the enemy may cause the high command to use precautionary tactics befitting an army low in confidence and reluctant to fight. A loud, cohesive and enthusiastic battle expression compared to the enemy's weak one may alert the military high command to undertake offensive, confident action to capitalise upon the eagerness of their own army and lack of enthusiasm on the part of the enemy.

The cohesive displays of vocal noise, instrumental sound, rhythmic movement and coordinated appearance that Graeco-Roman armies undertook on the battlefield reveals insight into military training regimes and battle preparedness. Prior learning and familiarity of religious hymns, poetry and other lyrical performances on the battlefield suggests that battle expression types were adopted from cultural life and integrated into the military practice. Instrumental sound (trumpets, drums, flutes) coordinated movement (rhythmic jumping, gesticulations) and intentional appearance undertaken by large numbers in unison reveals high levels of battle preparation and rehearsal for effective execution. The implementation of battle expression types into battlefield contexts suggests sophisticated battle preparedness on the part of military high command structures across the divergent cultural groups within the Graeco-Roman Mediterranean world. This has not been acknowledged by other studies.

Taunting of the enemy and spontaneous battle expression types showcase examples of culturally specific wit and humour used in a military context. Attempts to taunt the enemy and ridicule their behaviour or appearance served to instil greater levels of confidence amongst the agitators. On the battlefield, military forces used culturally specific wit and humour to unite along ethnic lines friendly troops and isolate as inferiors the enemy who, often, were culturally and ethnically alien. Through studying battlefield taunting and spontaneous acclamations from the literary record and archaeological sources, such as lead sling bullets, interesting features regarding culturally specific social customs, language use, sarcasm/satire/sadism and sexual connotations are understood.

By investigating why the battle expression was a typical feature of Graeco-Roman warfare, an understanding of the psychological dimension of battle in antiquity can be achieved. From the perspective of those waiting for battle in the front line of an ancient army, whether veteran or novice, it would have been a terrifying experience. The potential to suffer death or serious injury would have rendered many to unfamiliar levels of psychological wellbeing. The inherent instinct to feel protected and confident resulted in the fighting ranks of an army banding together, often along mutual and shared experience. Cohesive battle expression types were employed to positively divert the psychological state of a fighting force ready for combat. Fear and terror could render an army uncontrollable, whereas focused, confident armies were more passive. Pre-planned, rehearsed battle expression types served to familiarise the fighting ranks of their training and orders issued. Spontaneous battle expression types attempted to instil greater confidence within an army and impose thoughts of superiority over the enemy. The psychological state of an army could be gauged by the enthusiasm exhibited through undertaking a battle expression. Pre-battle atmospheric monomachy between opposing armies could determine, or influence, the outcome of a battle. The psychological state of an army could dictate to the high command what military strategy and tactics to be adopted to accommodate for this.

Besides the positive psychological factors that could be achieved through a battle expression, a key understanding of battlefield psychology is revealed when studying the objective to impose fear and intimidation over the enemy. The generation of an atmosphere hostile to the enemy was a key psychological factor in undertaking battle expression types. Cohesive, aggressive and imposing battle expression types served to inspire terror into the enemy ranks. The unfamiliar cultural traits evident in a battle expression aimed to unsettle the, often, culturally different enemy. Armies aimed to create sonic and visual displays to gain a psychological edge over the enemy before violent engagement took place. The Roman army designed their military training around exposing their soldiers to the alien customs of their enemies to reduce the devastating impacts of the enemy's battle expression. The religious sentiment behind many battle expression types served to focus the psychology of military forces against the potential dangers of terror from an enemy's battle expression.

The potential to associate the principles of the battle expression paradigm and to implement them for a study on *other* military cultures from geographical regions and historical periods outside of the Graeco-Roman Mediterranean world demonstrates its validity. For example, the study of cultural groups from the biblical Near East and the fertile crescent before the classical Graeco-Roman period, as well as, the study of post-Roman and medieval Europe would serve to highlight this. By comparing the sophistication, effectiveness and atmosphere generated by football supporters inside stadiums in Europe, the modern world can gain greater insight into the scale and range of an ancient military phenomenon that has since been misunderstood.

The validity of the battle expression paradigm from the Graeco-Roman world can be applied to later time periods and geographical regions. An 1855 study of "*War-Cries of Irish Septs*"[2] details what is known about the statements or words that surviving Irish clans cried in unison before, during and/or after battle for centuries. This study lists all known Irish clans whose war cry was recorded to have been heard on battlefields throughout "Tyrone's War"[3] otherwise referred to as Tyrone's Rebellion or most commonly the Nine Years War (1594–1603). In this list, 27 Irish clans have been named and next to each clan name was the Gaelic term or statement that its warriors cried out in battle.[4] The remainder of the study attempts to identify other known Irish clans and their respective war cries. The Gaelic terms listed are, on occasion, translated into English. The war cries of the Irish clans referred to different identifiable features that distinguished a clan from another. Specific character traits of the clan and its members were vocalised in battle, for example, the MacGilla Patricks raised the cry *Gear-laidir-aboo* which translates to "the sharp and strong".[5] Geographical landmarks, which describe the location of the clan's native land were also common, for example, the Tooles clan cried *Fear a-cnoic* "the men of the hill". The history of a tribe or honouring a former famous member was also common, for example, Clanrickarde's *Gall-ruadh* remembered the "Red foreigner" or "Red Earl" which may refer to the celebrated general Richard the Earl of Ulster.[6] Making known the clan name was another common feature of Irish clan battle cries, for example, the Barry clan would cry *Barragh-aboe* "a man of the Barrys". The knowledge of Irish battle war cries supports the notion that ancient Celtic/Germanic battle expressions took on a similar nature and range. There are comparable examples from the Graeco-Roman literary sources to the Irish war cries. For the most part, the Irish war cries, too, reflected natural elements associated with each clan. For example, the natural traits of clan members, the natural terrain of clan lands and famous individuals from the history of the clan are all reflective of the natural surroundings and substance of the Celtic/Germanic culture.

There is great potential for the battle expression paradigm to be superimposed onto contemporary as well as later time periods to the Graeco-Roman world from alternate geographical regions. Similarly, football supporter groups and the atmosphere they generate inside stadiums in the modern day can reveal insight into the sound, visuals and participation levels that were required in ancient times to create effective displays. Comparison and appreciation for the ancient military battle expression can be gauged by witnessing the atmosphere found in the practices of football supporter groups.

Ancient literary and archaeological evidence reveals that the battle expression was used to measure and provoke the psychological capacity of individuals and units that fought within armies on the battlefield. The incredible atmosphere of sight and sound that manifested from these performances tested the resolve of those exposed to it. A key aspect of the battle expression was the role it played in harnessing and boosting the collective fighting spirit of a military force that undertook it. Evidence suggests that military forces from the Graeco-Roman world focused on common cultural, religious and socio-political identifiers to galvanise their fighting force for battle. Moreover, similar identifiers could be utilised to taunt an enemy force on the battlefield to affect their psychological capacity at the onset of battle. The common identifiers that inspired battle expression forms provide valuable insight into the cultural, religious and socio-political features of societies in the Graeco-Roman world. It is interesting to understand what groups of men in a military context perceived as motivating factors to prepare themselves for violent confrontation with an enemy. Evidence reveals that the battle expression did play a role in contributing to the outcome of a battle and the decisions made by the military high command. For these reasons, the battle expression should be considered an important ancient military phenomenon.

Notes

1 As a specific example. South American football supporters also create similar types of intimidatory and impressive atmospheres as well.
2 Ulster Journal of Archaeology. "*War-Cries of Irish Septs*". Ulster Journal of Archaeology, First Series, Vol. 3 (1855), pp. 203–212. This journal article does not name an author. It is a dated source; however, its contents are significant in supporting the arguments here.
3 War-Cries (1855): 207.
4 War-Cries (1855): 207–208.
5 War-Cries (1855): 208 note "c".
6 War-Cries (1855): 207–208 note "a".

Bibliography

Aeschines. "*On the Embassy*", trans. Adams, C. Loeb Classical Library. Cambridge, MA: Harvard University Press, 2014.

Aeschylus. "*Agamemnon*", trans. Sommerstein, A. Loeb Classical Library. Cambridge, MA: Harvard University Press, 2014.

Aeschylus. "*Persians*", trans. Sommerstein, A. Loeb Classical Library. Cambridge, MA: Harvard University Press, 2014.

Aeschylus. "*Seven Against Thebes*", trans. Sommerstein, A. Loeb Classical Library. Cambridge, MA: Harvard University Press, 2014.

Alexandrescu, Cristina-Georgeta. "*The Iconography of Wind Instruments in Ancient Rome: Cornu, Bucina, Tuba, and Lituus*" Music in Art, Vol. 32, No. 1/2, Music in Art: Iconography as a Source for Music. History Volume III (Spring–Fall, 2007), pp. 33–46.

Ammianus Marcellinus. "*The Later Roman Empire (A.D. 352–378)*", trans. Hamilton, W. London: Penguin Classics, 1986.

Ammianus Marcellinus. "*Res Gestae*", trans. Rolfe, J. Loeb Classical Library. Cambridge, MA: Harvard University Press, 2014.

Appian. "*Civil Wars*", trans. White, H. Loeb Classical Library. Cambridge, MA: Harvard University Press, 2014.

Ariño, Borja Diaz. "Glandes inscriptae de la Península Ibérica" *Zeitschrift für Papyrologie und Epigraphik*, Bd. 153 (2005), pp. 219–236.

Aristophanes. "*Acharnians*", trans. Henderson, J. Loeb Classical Library. Cambridge, MA: Harvard University Press, 2014.

Aristophanes. "*Birds*", trans. Henderson, J. Loeb Classical Library. Cambridge, MA: Harvard University Press, 2014.

Aristophanes. "*Knights*", trans. Henderson, J. Loeb Classical Library. Cambridge, MA: Harvard University Press, 2014.

Aristophanes. "*Peace*", trans. Henderson, J. Loeb Classical Library. Cambridge, MA: Harvard University Press, 2014.

Armstrong, Jeremy. "*Claiming Victory: The Early Roman Triumph*" in Rituals of Triumph in the Mediterranean World. Eds: Anthony Spalinger and Jeremy Armstrong. Vol. 63 Culture and History of the Ancient Near East. Leiden; Boston, 2013, pp. 7–21.

Arrian. "*Anabasis*", trans. Brunt, P. Loeb Classical Library. Cambridge, MA: Harvard University Press, 2014.

Arrian. "*The Campaigns of Alexander*", trans. dev Selincourt, A. Revised by Hamilton, J. R. London: Penguin Classics, 1971.

Asheri, David. "*Book I*" in A Commentary on Herodotus Books I-IV edited by: Oswyn Murray and Alfonso Moreno; with a contribution by: Maria Brosius; translated by: Barbara Graziosi . . . [et al.]. Oxford: Oxford University Press, 2007, pp. 57–218.

Athenaeus. *"Deipnosophistae"*, trans. Gulick, C. Loeb Classical Library. Cambridge, MA: Harvard University Press, 2014.

Badian, Ernst. *"Arrianus"* in Brill's New Pauly, Antiquity volumes edited by: Hubert Cancik and Helmuth Schneider, English Edition by: Christine F. Salazar, Classical Tradition volumes edited by: Manfred Landfester, English Edition by: Francis G. Gentry. Leiden; Boston: Brill, 2002.

Balot, Ryan. *"Courage in the Democratic Polis"* The Classical Quarterly, New Series, Vol. 54, No. 2 (Dec., 2004), pp. 406–423.

Banks, J. *The Idylls of Theocritus, Bion, and Moschus, and the War-Songs of Tyrtaeus.* London, 1853.

Baray, Luc. *"Celtes, Galates et Gaulois. Mercenaires de l'Antiquité. Representation, recrutement, organization"* Paris: Editions Picard, 2017.

Baray, Luc. *"Les mercenaires celtes et la culture de La Tène: critères archéologiques et positions sociologiques"* Dijon: Editions universitaires de Dijon, 2014.

Barrett, Anthony A. *"Review: [untitled]"* The Classical World, Vol. 80, No. 4 (Mar.–Apr., 1987), p. 323: Classical Association of the Atlantic States.

Bayliss, Andrew. *"Good to Slaughter the Lives of Young Men? The Role of Tyrtaeus' Poetry in Spartan Society"* Esparta: Politica e Sociedade/Luiz Filipe Bantim de Assumpcao, 1st edition. Curitiba: Editora Prismas, 2017, pp. 49–86.

Baynham, Elizabeth. *"Barbarians I: Quintus Curtius' and Other Roman Historians' Reception of Alexander"* in Chapter 18: The Cambridge Companion to the Roman Historians. Ed: Feldherr, A. Cambridge: Cambridge University Press, 2009, pp. 288–300.

Beard, Mary. *"The Roman Triumph"* Cambridge, MA: Belknap Press of Harvard University Press, 2007.

Birley, A. R. *"Cassius"* in Brill's New Pauly, Antiquity volumes edited by: Hubert Cancik and Helmuth Schneider, English Edition by: Christine F. Salazar, Classical Tradition volumes edited by: Manfred Landfester, English Edition by: Francis G. Gentry. Leiden; Boston: Brill, 2002.

Blench, R. *"Archaeology, Language and the African Past"* Lanham, MD: Rowman Altamira, 2006.

Bowie, Ewen. *"Athenaeus"* in Brill's New Pauly, Antiquity volumes edited by: Hubert Cancik and, Helmuth Schneider, English Edition by: Christine F. Salazar, Classical Tradition volumes edited by: Manfred Landfester, English Edition by: Francis G. Gentry. Leiden; Boston: Brill, 2002.

Bowie, Ewen. *"Tyrtaeus"* in Brill's New Pauly, Antiquity volumes edited by: Hubert Cancik and, Helmuth Schneider, English Edition by: Christine F. Salazar, Classical Tradition volumes edited by: Manfred Landfester, English Edition by: Francis G. Gentry. Leiden; Boston: Brill, 2002.

Brandt, Hartwin. *"Vegetius"* in Brill's New Pauly, Antiquity volumes edited by: Hubert Cancik and, Helmuth Schneider, English Edition by: Christine F. Salazar, Classical Tradition volumes edited by: Manfred Landfester, English Edition by: Francis G. Gentry. Leiden; Boston: Brill, 2002.

Brilliant, Richard. *"'Let the Trumpets Roar!' The Roman Triumph"* Studies in the History of Art, Vol. 56, Symposium Papers XXXIV: The Art of Ancient Spectacle (1999), pp. 220–229.

Bruun, Patrick. *"The Battle of the Milvian Bridge: The Date Reconsidered"* Hermes, 88. Bd., H. 3 (Jul., 1960), pp. 361–370.

Bruun, Patrick. *"The Victorious Signs of Constantine: A Reappraisal"* The Numismatic Chronicle (1966–), Vol. 157 (1997), pp. 41–59.

Buck, Carl D. *"Words for 'Battle,' 'War,' 'Army,' and 'Soldier'"* Classical Philology, Vol. 14, No. 1 (Jan., 1919), pp. 1–19.

Caesar. "African War", trans. Loeb Classical Library. Cambridge, MA: Harvard University Press, 2014.

Caesar. *"Civil Wars"*, trans. Damon, C. Loeb Classical Library. Cambridge, MA: Harvard University Press, 2014.

Caesar. *"Gallic War"*, trans. Edwards, H. Loeb Classical Library. Cambridge, MA: Harvard University Press, 2014.

Caesar. *"Spanish War"*, trans. Way, A. Loeb Classical Library. Cambridge, MA: Harvard University Press, 2014.

Cameron, Alan. *"Notes on Claudian's Invectives"* The Classical Quarterly, New Series, Vol. 18, No. 2 (Nov., 1968), pp. 387–411.

Campbell, Brian. *"The Roman Army, 31 BC-AD 337: A Sourcebook"* London: Routledge, 1994.

Campbell, J. B. *"The Emperor and the Roman Army, 31 BC-AD 235"* Oxford: Oxford University Press, 1984.

Campbell, J. B. *"Mutiny"* in Brill's New Pauly, Antiquity volumes edited by: Hubert Cancik and Helmuth Schneider, English Edition by: Christine F. Salazar, Classical Tradition volumes edited by: Manfred Landfester, English Edition by: Francis G. Gentry. Leiden; Boston: Brill, 2002.

Cartledge, Paul. *"The Spartans: The World of the Warrior-Heroes of Ancient Greece, from Utopia to Crisis and Collapse"* Woodstock, NY: Overlook Press, 2003.

Cartledge, Paul. *"Thermopylae: The Battle That Changed the World"* London: Macmillan, 2006.

Connolly, Peter. *"The Roman Army"* London: Macdonald Educational, 1975.

Cowan, Ross. *"The Clashing of Weapons and Silent Advances in Roman Battles"* Historia: Zeitschrift fur Alte Geschichte, Bd. 56, H. 1 (2007), pp. 114–117.

Crowley, Jason. *"Beyond the Universal Soldier: Combat Trauma in Classical Antiquity"* Combat Trauma and the Ancient Greeks. Eds: Meineck, P. and Konstan, D. London: Palgrave Macmillan, 2014, pp. 105–130.

Crowley, Jason. *"The Psychology of the Athenian Hoplite"* Cambridge: Cambridge University Press, 2012.

Dallmann, Volker. *"Homerus (Homer)"* in Brill's New Pauly Supplements I – Volume 2: Dictionary of Greek and Latin Authors and Texts edited by: Manfred Landfester, in collaboration with Brigitte Egger. Leiden; Boston: Brill, 2002.

Davies, Roy W. *"Service in the Roman Army"* edited by David Breeze and Valerie A. Maxfield. Edinburgh, UK: Edinburgh University Press with the Publications Board of the University of Durham, 1989.

De Vivo, Juan Sebastian. *"The Memory of Greek Battle: Material Culture and/as Narrative of Combat"* Combat Trauma and the Ancient Greeks. Eds" Meineck, P. and Konstan, D. London: Palgrave Macmillan, 2014, pp. 163–184.

Dio Cassius. *"Roman History"*, trans. Cary, E. and Foster, H. Loeb Classical Library. Cambridge, MA: Harvard University Press, 2014.

Diodorus Siculus. *"The Library of History"*, trans. Oldfather, C. Loeb Classical Library. Cambridge, MA: Harvard University Press, 2014.

Dionysius of Halicarnassus. *"Roman Antiquities"*, trans. Cary, E. Loeb Classical Library. Cambridge, MA: Harvard University Press, 2014.

Dobson, B. and Mann, J. C. *"The Roman Army in Britain and Britons in the Roman Army"* Britannia, Vol. 4 (1973), pp. 191–205.

Dreyer, Boris. *"Polybius"* in Brill's New Pauly, Antiquity volumes edited by: Hubert Cancik and Helmuth Schneider, English Edition by: Christine F. Salazar, Classical Tradition volumes edited by: Manfred Landfester, English Edition by: Francis G. Gentry. Leiden; Boston: Brill, 2002.

Dusanic, Slobodan. *"The Imperial Propaganda of Significant Day-Dates: Two Notes in Military History"* Bulletin of the Institute of Classical Studies. Supplement, No. 81, Documenting the Roman Army: Essays in Honour of Margaret Roxan. Hoboken, NJ: Wiley, 2003, pp. 89–100.

Eder, Walter. *"Triumph, Triumphal Procession"* in Brill's New Pauly, Antiquity volumes edited by: Hubert Cancik and Helmuth Schneider, English Edition by: Christine F. Salazar, Classical Tradition volumes edited by: Manfred Landfester, English Edition by: Francis G. Gentry. Leiden; Boston: Brill, 2006.

Elton, Hugh. *"Warfare in Roman Europe, AD 350–425"* Oxford: Oxford University Press, 1996.

Ephraim, David. *"Sparta's Kosmos of Silence"* edited by: Stephen Hodkinson and Anton Powell Sparta: New Perspectives; Swansea: Classical Press of Wales, 1999, pp. 117–146.

Erskine, Andrew. *"Hellenistic Parades and Roman Triumphs"* in Rituals of Triumph in the Mediterranean World. Eds: Anthony Spalinger and Jeremy Armstrong. Vol. 63, Culture and History of the Ancient Near East. Leiden, Boston: Brill, 2013, pp. 37–55.

Eusebius. *"Life of Constantine"*, trans. Cameron, A. and Hall, S. Oxford: Clarendon Press, 1999.

Faber, Riemer. *"Vergil's 'Shield of Aeneas' ('Aeneid' 8. 617–731) and the 'Shield of Heracles'"* Mnemosyne, Fourth Series, Vol. 53, Fasc. 1 (Feb., 2000), pp. 49–57.

Feldherr, Andrew. *"Barbarians II: Tacitus' Jews"* in Chapter 19: The Cambridge Companion to the Roman Historians. Ed. Feldherr, A. Cambridge: Cambridge University Press, 2009, pp. 301–316.

Flaig, Egon. *"Tacitus"* in Brill's New Pauly, Antiquity volumes edited by: Hubert Cancik and, Helmuth Schneider, English Edition by: Christine F. Salazar, Classical Tradition volumes edited by: Manfred Landfester, English Edition by: Francis G. Gentry. Leiden; Boston: Brill, 2002.

Flower, Michael. *"Spartan Religion"* in A Companion to Sparta. Ed: Powell, A. Malden MA: Wiley Blackwell, 2017, pp. 425–451.

Ford, Andrew. *"The Genre of Genres: Paeans and Paian in Early Greek Poetry"* Poetica, Vol. 38, No. 3/4 (2006), pp. 277–295.

Foss, Clive. *"A Bullet of Tissaphernes"* The Journal of Hellenic Studies, Vol. 95 (1975), pp. 25–30.

Foss, Clive. *"Greek Sling Bullets in Oxford"* Archaeological Reports, No. 21 (1974–1975) pp. 40–44.

Frank, Tenney. *"The Diplomacy of Q. Marcius in 169 B. C."* Classical Philology, Vol. 5, No. 3 (Jul., 1910), pp. 358–361.

Frank, Tenney. *"Roman Historiography before Caesar"* The American Historical Review, Vol. 32, No. 2 (Jan., 1927), pp. 232–240.

Frontinus. *"Strategems"*, trans. Bennett, C. Loeb Classical Library. Cambridge, MA: Harvard University Press, 2014.

Furley, William D. *"Praise and Persuasion in Greek Hymns"* The Journal of Hellenic Studies, Vol. 115 (1995), pp. 29–46: The Society for the Promotion of Hellenic Studies.

Gilliam, J. F. *"Roman Army"* Amsterdam: J.C. Gieben, 1986.

Gilliver, Catherine M. *"Battle"* in Chapter 4: The Cambridge History of Greek and Roman Warfare Volume 2: Rome from the Late Republic to the Late Empire. Eds: Philip Sabin, King's College London, Hans van Wees, Michael Whitby. Cambridge: Cambridge University Press, 2007, pp. 122–157.

Gleason, Bruce. "*Cavalry Trumpet and Kettledrum Practice From the Time of the Celts and Romans to the Renaissance*" The Galpin Society Journal, Vol. 61 (Apr., 2008), pp. 231–239, 251.

Glück, J. J. "*Reviling and Monomachy as Battle-Preludes in Ancient Warfare*" Acta Classica, Vol. 7. Classical Association of South Africa (1964), pp. 25–31.

Goldsworthy, Adrian. "*The Complete Roman Army*" London: Thames & Hudson, 2003.

Goldsworthy, Adrian. "*The Roman Army at War: 100 BC-AD 200*" Oxford: Clarendon Press, 1996.

Grancsay, Stephen V. "*A Fifteenth-Century Painted Shield*" The Metropolitan Museum of Art Bulletin, Vol. 26, No. 1 (Jan., 1931), pp. 12–14: The Metropolitan Museum of Art.

Grant, Michael. "*The army of the Caesars*" London: Weidenfeld & Nicolson, 1974.

Grant, Michael. "*Greek and Roman Historians: Information and Misinformation*" London: Routledge, 1995.

Haldane, J. A. "*Musical Themes and Imagery in Aeschylus*" The Journal of Hellenic Studies, Vol. 85 (1965), pp. 33–41.

Hallett, Judith. "*Perusinae Glandes and the Changing Image of Augustus*" AJAH, Vol. 2 (1977), pp. 151–171.

Hammond, Nicholas G. L. "*Alexander's Charge at the Battle of Issus in 333 B.C.*" Historia: Zeitschrift für Alte Geschichte, Bd. 41, H. 4 (1992), pp. 395–406.

Hanson, Victor Davis. "*The Western Way of War*" Oxford: Oxford University Press, 1989.

Harris, Stephen and Platzner, Gloria. "*Classical Mythology: Images and Insights*" 3rd edition. California City, CA: Mayfield Publishing Company, 2001.

Harrison, S. J. "*The Survival and Supremacy of Rome: The Unity of the Shield of Aeneas*" The Journal of Roman Studies, Vol. 87 (1997), pp. 70–76.

Haynes, Ian. "*Blood of the Provinces: The Roman Auxilia and the Making of Provincial Society From Augustus to the Severans*" Oxford: Oxford University Press, 2013.

Hempl, George. "*The Salian Hymn to Janus*" Transactions and Proceedings of the American Philological Association, Vol. 31 (1900), pp. 182–188.

Herodotus. "*The Persian Wars*", trans. Godley, A. Loeb Classical Library. Cambridge, MA: Harvard University Press, 2014.

HO, S. "*Thucydides*" in Brill's New Pauly, Antiquity volumes edited by: Hubert Cancik and Helmuth Schneider, English Edition by: Christine F. Salazar, Classical Tradition volumes edited by: Manfred Landfester, English Edition by: Francis G. Gentry. Leiden; Boston: Brill, 2002.

Höcker, Christoph. "*Monumental Columns*" in Brill's New Pauly, Antiquity volumes edited by: Hubert Cancik and, Helmuth Schneider, English Edition by: Christine F. Salazar, Classical Tradition volumes edited by: Manfred Landfester, English Edition by: Francis G. Gentry. Leiden; Boston: Brill, 2002.

Hodkinson, Stephen and Powell, Anton, eds. "*Sparta & War*" Swansea: Classical Press of Wales, 2006.

Hoesch, Nicola. "*Alexander Mosaic*" in Brill's New Pauly, Antiquity volumes edited by: Hubert Cancik and, Helmuth Schneider, English Edition by: Christine F. Salazar, Classical Tradition volumes edited by: Manfred Landfester, English Edition by: Francis G. Gentry. Leiden; Boston: Brill, 2002.

Hölter, Achim. "*Homer*" in Brill's New Pauly Supplements II – Volume 7: Figures of Antiquity and Their Reception in Art, Literature and Music, English edition by Chad M. Schroeder. Leiden; Boston: Brill, 2016.

Homer. "*Iliad*", trans. Murray, A. Loeb Classical Library. Cambridge, MA: Harvard University Press, 2014.

Homeric Hymns 3. *"To Apollo"*, trans. West, M. Loeb Classical Library. Cambridge, MA: Harvard University Press, 2014.

Homeric Hymns 8. *"To Ares"*, trans. West, M. Loeb Classical Library. Cambridge, MA: Harvard University Press, 2014.

Hornblower, S., Spawforth, A. and Eidinow, E., eds. *"The Oxford Classical Dictionary"*, 4th edition. Oxford: Oxford University Press, 2012.

How, W. W. and Wells, J. *"A Commentary on Herodotus"* Project Gutenberg EBook, 2008.

Ingold, T., ed. *"Companion Encyclopedia of Anthropology"* Abingdon, UK: Routledge, 2002.

Josephus. *"Jewish War"*, trans. Thackeray, H. Loeb Classical Library. Cambridge, MA: Harvard University Press, 2014.

Keegan, John. *"The Face of Battle"* London: Cape, 1977.

Keith, Alison M. *"Ovid on Vergilian War Narrative"* Vergilius (1959–), Vol. 48. Bacoli, Italy: The Vergilian Society, 2002, pp. 105–122.

Kellett, Anthony. *"The Soldier in Battle: Motivational and Behavioral Aspects of the Combat Experience"* in From Psychological Dimensions of War. Eds: Betty Glad. Newberry Park, CA: Sage Publications, 1990, pp. 215–235.

Kelly, Amanda. *"The Cretan Slinger at War: A Weighty Exchange"* The Annual of the British School at Athens, Vol. 107 (Nov. 2012), pp. 273–311.

Kelly, Gavin. *"Ammianus Marcellinus: Tacitus' Heir and Gibbon's Guide"* The Roman Historians. Ed: Andrew Feldherr. Cambridge: Cambridge University Press, 2009, pp. 348–361.

Kinsella, Thomas, trans. *"The Táin"* Oxford: Oxford University Press, 1969, pp. 150–153.

Kirk, G. S. *"The Iliad: A Commentary: Vol 1 Books 1–4"* Cambridge: Cambridge University Press, 1985.

Kirk, G. S. *"The Iliad: A Commentary: Vol 2 Books 5–8"* Cambridge: Cambridge University Press, 1990.

Koepke, Nikola and Baten, Joerg. *"The Biological Standard of Living in Europe during the Last Two Millennia"* European Review of Economic History, Vol. 9, No. 1 (Apr., 2005), pp. 61–95.

Kõiv, Mait. *"The Origins, Development, and Reliability of the Ancient Tradition About the Formation of the Spartan Constitution"* Historia (Wiesbaden, Germany), Vol. 54 (Jan. 2005), pp. 233–264.

Krentz, Peter. *"The Nature of Hoplite Battle"* Classical Antiquity, Vol. 4, No. 1 (Apr., 1985), pp. 50–61.

Krentz, Peter. *"The Oath of Marathon, Not Plataia?"* Hesperia: The Journal of the American School of Classical Studies at Athens. The American School of Classical Studies at Athens, Vol. 76, No. 4 (Oct.–Dec., 2007), pp. 731–742.

Lactantius. *"De Mortibus Persecutorum"*, trans. Creed, J. L. Oxford: Clarendon Press, 1984.

Lanni, Adrian. *"The Laws of War in Ancient Greece"* Law and History Review, Vol. 26, No. 3, Law, War, and History (Fall, 2008), pp. 469–489.

Lazenby, J. F. *"The Spartan Army"* Barnsley: Pen & Sword Military, 2012.

Lendon, J. E. *"The Rhetoric of Combat: Greek Military Theory and Roman Culture in Julius Caesar's Battle Descriptions"* Classical Antiquity, Vol. 18, No. 2. Oakland: University of California Press, Oct., 1999, pp. 273–329.

Lewis, Charlton and Short, Charles. *"A Latin Dictionary"* Oxford: Clarendon Press, 1879.

Liddell, Henry and Scott, Robert. *"A Greek-English Lexicon"* Oxford: Clarendon Press, 1940.

Lindow, John. *"Norse Mythology: A Guide to the Gods, Heroes, Rituals, and Beliefs"* Oxford: Oxford University Press, 2001.

Littleton, C. S. and Thomas, A. C. *"The Sarmatian Connection: New Light on the Origin of the Arthurian and Holy Grail Legends"* The Journal of American Folklore, Vol. 91, No. 359 (Jan.–Mar., 1978), pp. 513–527.

Livy. *"History of Rome: Books 3–23"*, trans. Foster, B. Loeb Classical Library. Cambridge, MA: Harvard University Press, 2014.

Livy. *"History of Rome: Books 24–30"*, trans. Moore, F. Loeb Classical Library. Cambridge, MA: Harvard University Press, 2017.

Livy. *"History of Rome: Books 32–37"*, trans. Yardley, J. Loeb Classical Library. Cambridge, MA: Harvard University Press, 2017.

Livy. *"History of Rome: Books 38"*, trans. Sage, E. Loeb Classical Library. Cambridge, MA: Harvard University Press, 2017.

Livy. *"History of Rome: Books 45"*, trans. Schlesinger, C. Loeb Classical Library. Cambridge, MA: Harvard University Press, 2014.

Lloyd-Morgan, G. *"Nemesis and Bellona: A Preliminary Study of Two Neglected Goddesses"* in The Concept of the Goddess Abingdon, UK: Routledge, 1996, pp. 125–126.

Lobell, Jarrett and Patel, Samir. *"Bog Bodies Rediscovered"* Archaeology, Vol. 63, No. 3 (May/June 2010), pp. 22–29.

Lucan. *"Pharsalia (The Civil War)"*, trans. Duff, J. D. Loeb Classical Library. Cambridge, MA: Harvard University Press, 2014.

Lysias. *"In Defence of Mantitheus"*, trans. Lamb, W. Loeb Classical Library. Cambridge, MA: Harvard University Press, 2014.

Macdonald, Fiona. *"Roman Soldier"* London: Pan Macmillan, 1992.

Marincola, John. *"Ancient Audiences and Expectations"* in Chapter 1: The Cambridge Companion to the Roman Historians. Ed. Feldherr, A. Cambridge: Cambridge University Press, 2009, pp. 11–23.

Master Wace. *"His Chronicle of the Norman Conquest from the Roman de Rou"* trans. Taylor, Edgar. London: W. Pickering, 1837.

Maurice. *"Strategikon"*, trans. Dennis, G. Philadelphia: University of Pennsylvania Press, 2001.

McDermott, William C. *"Glandes Plumbeae"* The Classical Journal, Vol. 38, No. 1 (Oct., 1942), pp. 35–37.

McWhiney, Grady and Jamieson, Perry D. *"Attack and Die: Civil War Military Tactics and the Southern Heritage"* Tuscaloosa: University Alabama Press, 1984.

Meister, Klaus. *"Herodotus"* in Brill's New Pauly, Antiquity volumes edited by: Hubert Cancik and, Helmuth Schneider, English Edition by: Christine F. Salazar, Classical Tradition volumes edited by: Manfred Landfester, English Edition by: Francis G. Gentry. Leiden; Boston: Brill, 2002.

Meister, Klaus. *"Polyaenus"* in Brill's New Pauly, Antiquity volumes edited by: Hubert Cancik and, Helmuth Schneider, English Edition by: Christine F. Salazar, Classical Tradition volumes edited by: Manfred Landfester, English Edition by: Francis G. Gentry. Leiden; Boston: Brill, 2002.

Meucci, Renato. *"Roman Military Instruments and the Lituus"* The Galpin Society Journal, Vol. 42 (Aug., 1989), pp. 85–97.

Moosavi, Amir. *"How to Write Death: Resignifying Martyrdom in Two Novels of the Iran-Iraq War"* Alif: Journal of Comparative Poetics, No. 35, New Paradigms in the Study of Middle Eastern Literatures (2015), pp. 9–31.

Nesselrath, Heinz-Günther. "*Aristophanes*" in Brill's New Pauly, Antiquity volumes edited by: Hubert Cancik and Helmuth Schneider, English Edition by: Christine F. Salazar, Classical Tradition volumes edited by: Manfred Landfester, English Edition by: Francis G. Gentry. Leiden; Boston: Brill, 2002.

Onosander. "*Strategikos*", trans. Illinois Greek Club Loeb Classical Library. Cambridge, MA: Harvard University Press, 2014.

Ovid. "*Amores*", trans. Showerman, G. Loeb Classical Library. Cambridge, MA: Harvard University Press, 2014.

Palaima, Thomas. "*When War Is Performed, What Do Soldiers and Veterans Want to Hear and See and Why?*" in Combat Trauma and the Ancient Greeks. Eds: Peter Meineck and David Konstan. New York: Palgrave Macmillan, 2014, pp. 261–285.

Perrin, Porter G. "*'Hubba Hubba' Scholium VI*" American Speech, Vol. 32, No. 3 (Oct., 1957), pp. 237–238: Duke University Press.

Pitcher, Luke. "*The Roman Historians After Livy*" in A Companion to Julius Caesar. Ed: Liriam Griffin. Hoboken, NJ: Wiley-Blackwell Publishing, 2009, pp. 267–276.

Plutarch. "*Aemilius Paulus*", trans. Perrin, B. Loeb Classical Library. Cambridge, MA: Harvard University Press, 2014.

Plutarch. "*Alexander*", trans. Perrin, B. Loeb Classical Library. Cambridge, MA: Harvard University Press, 2014.

Plutarch. "*Antony*", trans. Perrin, B. Loeb Classical Library. Cambridge, MA: Harvard University Press, 2014.

Plutarch. "*Cleomenes*", trans. Perrin, B. Loeb Classical Library. Cambridge, MA: Harvard University Press, 2014.

Plutarch. "*Crassus*", trans. Perrin, B. Loeb Classical Library. Cambridge, MA: Harvard University Press, 2014.

Plutarch. "*Instituta Laconica*", trans. Babbitt, F. Loeb Classical Library. Cambridge, MA: Harvard University Press, 2014.

Plutarch. "*Lycurgus*", trans. Perrin, B. Loeb Classical Library. Cambridge, MA: Harvard University Press, 2014.

Plutarch. "*Marcellus*", trans. Perrin, B. Loeb Classical Library. Cambridge, MA: Harvard University Press, 2014.

Plutarch. "*Marius*", trans. Perrin, B. Loeb Classical Library. Cambridge, MA: Harvard University Press, 2014.

Plutarch. "*M. Cato*", trans. Perrin, B. Loeb Classical Library. Cambridge, MA: Harvard University Press, 2014.

Plutarch. "*Moralia*", trans. Cherniss and Helmbold Loeb Classical Library. Cambridge, MA: Harvard University Press, 2014.

Plutarch. "*Moralia, Sayings of Spartans (Apophthegmata Laconica)*", trans. Babbitt, F. Loeb Classical Library. Cambridge, MA: Harvard University Press, 2014.

Plutarch. "*Numa*", trans. Perrin, B. Loeb Classical Library. Cambridge, MA: Harvard University Press, 2014.

Plutarch. "*Pompey*", trans. Perrin, B. Loeb Classical Library. Cambridge, MA: Harvard University Press, 2014.

Plutarch. "*Sulla*", trans. Perrin, B. Loeb Classical Library. Cambridge, MA: Harvard University Press, 2014.

Polyaenus. "*Polyaenus's Stratagems of War (1793)*" trans. R. Shepherd. Whitefish, MT: Kessinger Publishing, 2010.

Polybius. "*The Histories*", trans. Paton, W. Loeb Classical Library. Cambridge, MA: Harvard University Press, 2014.

Potter, John. *"The Antiquities of Greece"* Printed for J. Knapton, R. Knaplock, J. and B. Sprint, D. Midwinter, A. Bettesworth, R. Robinson, W. and J. Innys, J. Osborne, T. Longman, W. Mears, and A. Ward, London, 1728.

Powell, Anton. *"The Women of Sparta: And of Other Greek Cities: At War"* in Spartan Society. Ed: T.J. Figueira Swansea: Classical Press of Wales, 2004, pp. 137–150.

Pritchett, W. Kendrick. *"The Greek State at War: Part I"* Berkeley: University of California Press, 1971.

Procopius. *"Wars"*, trans. Dewing, H. Loeb Classical Library. Cambridge, MA: Harvard University Press, 2014.

Pushkin, Alexander and Elton, Oliver. *"A Battle"* The Slavonic and East European Review, Vol. 12, No. 35 (Jan., 1934), p. 272: The Modern Humanities Research Association and University College London, School of Slavonic and East European Studies.

Quintus Curtius. *"History of Alexander"*, trans. Rolfe, J. Loeb Classical Library. Cambridge, MA: Harvard University Press, 2014.

Ramage, Edwin S. *"Demigration of Predecessor Under Claudius, Galba, and Vespasian"* Historia: Zeitschrift fur Alte Geschichte, Bd. 32, H.2, 2nd Qtr., 1983, pp. 201–214.

Rance, Philip. *"Battle"* in Chapter 10: The Cambridge History of Greek and Roman Warfare Volume 2: Rome from the Late Republic to the Late Empire. Eds: Philip Sabin, King's College London, Hans van Wees and Michael Whitby Cambridge: Cambridge University Press, 2007, pp. 342–378.

Rance, Philip. *"War Cry"* in The Encyclopedia of the Roman Army, 1st edition. Ed: Le Bohec, Y. Oxford: John Wiley and Sons Ltd., 2015, pp. 1–2.

Rawlinson, C. "On the Ethnography of the Cimbri." *The Journal of the Anthropological Institute of Great Britain and Ireland*, Vol. 6 (1877), pp. 150–158.

Richer, Nicola. *"La religion des Spartiates: croyances et cultes dans l'Antiquité"* Paris: Le Belles Lettres, 2012.

Richer, Nicola. *"The Religious System at Sparta"* in A Companion to Greek Religion. Ed: Ogden, D. Malden, MA: Wiley Blackwell, 2010, pp. 236–252.

Rives, J. B. *"Germania/Tacitus: Translated With Introduction and Commentary"* Oxford: Clarendon Press, 1999.

Rosen, Klaus. *"Ammianus Marcellinus"* in Brill's New Pauly, Antiquity volumes edited by: Hubert Cancik and Helmuth Schneider, English Edition by: Christine F. Salazar, Classical Tradition volumes edited by: Manfred Landfester, English Edition by: Francis G. Gentry. Leiden; Boston: Brill, 2002.

Rusch, Scott M. *"Sparta at War: Strateg, Tactics, and Campains, 550–362 BC"* Barnsley: Frontline Books, 2011.

Rutherford, Ian. *"Apollo in Ivy: The Tragic Paean"* Arion, Third Series, Vol. 3, No. 1, The Chorus in Greek Tragedy and Culture, One (Fall, 1994–Winter, 1995), pp. 112–135.

Rutherford, Ian. *"Neoptolemus and the Paean-Cry: An Echo of a Sacred Aetiology in Pindar"* Zeitschrift fur Papyrologie und Epigraphik, Bd. 88 (1991), pp. 1–10.

Sabin, Philip. *"The Face of Roman Battle"* The Journal of Roman Studies, Vol. 90 (2000), pp. 1–17.

Sabin, Philip and de Souza, P. *"Battle"* in Chapter 13: The Cambridge History of Greek and Roman Warfare: Volume 1: Greece, The Hellenistic World and the Rise of Rome. Eds: Sabin, P., van Wees, H. and Whitby, M. Cambridge: Cambridge University Press, 2007, pp. 399–460.

Sallmann, Klaus. *"Suetonius"* in Brill's New Pauly, Antiquity volumes edited by: Hubert Cancik and, Helmuth Schneider, English Edition by: Christine F. Salazar, Classical

Tradition volumes edited by: Manfred Landfester, English Edition by: Francis G. Gentry. Leiden; Boston: Brill, 2002.

Sallust. "*Jugurthine War*", trans. Rolfe, J. Loeb Classical Library. Cambridge, MA: Harvard University Press, 2014.

Seaford, Richard. "*The Attribution of Aeschylus, Choephoroi 691–9*" The Classical Quarterly, New Series, Vol. 39, No. 2 (1989), pp. 302–306.

Shelton, Jo-Ann. "*As the Romans Did: A Sourcebook in Roman Social History*" Oxford: Oxford University Press, 1988.

Simek, Rudolf. "*Dictionary of Northern Mythology*", trans. Angela Hall. Rochester, NY: Boydell & Brewer Ltd, 2007.

Sommerstein, Alan H. and Bayliss, Andrew J. "*Oath and State in Ancient Greece*" Berlin: De Gruyter, 2017.

Speidel, M. "*Ancient Germanic Warriors: Warrior Styles From Trajan's Column to Icelandic Sagas*" London: Routledge, 2004.

Speidel, M. "*Beserks: A History of Indo-European 'Mad Warriors* '" Journal of World History, Vol. 13, No. 2 (Fall, 2002), pp. 253–290.

Speidel, M. "*Mithras-Orion: Greek Hero and Roman Army God*" Leiden: Brill, 1980.

Speidel, M. "*The Religion of Iuppiter Dolichenus in the Roman Army*" Leiden: Brill, 1978.

Speidel, M. "*Riding for Caesar: The Roman Emperors' Horse Guards*" London: Batsford, 1994.

Speidel, M. "*Roman Army Studies*" Amsterdam: J.C. Gieben, 1984.

Spence, I. "*Historical Dictionary of Ancient Greek Warfare*" Lanham, MD: Scarecrow Press, 2002.

Spier, Jeffrey. "*Emblems in Archaic Greece*" Bulletin of the Institute of Classical Studies, No. 37 (1990), pp. 107–129. 114 BICS 37.

Strabo. "*Geography*", trans. Jones, H. Loeb Classical Library. Cambridge, MA: Harvard University Press, 2014.

Suetonius. "*Twelve Caesars*", trans. Rolfe, J. Loeb Classical Library. Cambridge, MA: Harvard University Press, 2014.

Tacitus. "*Annals*", trans. Jackson, J. Loeb Classical Library. Cambridge, MA: Harvard University Press, 2014.

Tacitus. "*Germania*", trans. Hutton and Peterson. Loeb Classical Library. Cambridge, MA: Harvard University Press, 2014.

Tacitus. "*Histories*", trans. Moore, C. Loeb Classical Library. Cambridge, MA: Harvard University Press, 2014.

Thucydides. "*Peloponnesian War*", trans. Smith, C. Loeb Classical Library. Cambridge, MA: Harvard University Press, 2014.

Tinnefeld, Franz. "*Mauricius*" in Brill's New Pauly, Antiquity volumes edited by: Hubert Cancik and, Helmuth Schneider, English Edition by: Christine F. Salazar, Classical Tradition volumes edited by: Manfred Landfester, English Edition by: Francis G. Gentry. Leiden; Boston: Brill, 2002.

Tinnefeld, Franz. "*Procopius*" in Brill's New Pauly, Antiquity volumes edited by: Hubert Cancik and, Helmuth Schneider, English Edition by: Christine F. Salazar, Classical Tradition volumes edited by: Manfred Landfester, English Edition by: Francis G. Gentry. Leiden; Boston: Brill, 2002.

Torrance, Isabelle. "*Distorted Oaths in Aeschylus*" Illinois Classical Studies, Vol. 40, No. 2 (Fall, 2015), pp. 281–295.

Tyrtaeus. "*The Idylls of Theocritus, Bion and Moschus; and the War-Songs of Tyrtaeus*", trans. Banks, J. London: H. G. Bohn, 1853.

Ulster Journal of Archaeology. *"War-Cries of Irish Septs"* Ulster Journal of Archaeology, First Series, Vol. 3 (1855), pp. 203–212.

Valerius Maximus. *"Memorable Doings and Sayings"*, trans. Shackleton Bailey, D. Loeb Classical Library. Cambridge, MA: Harvard University Press, 2014.

Van Wees, Hans. *"The Oath of the Sworn Bands: The Acharnae Stela, the Oath of Plataea and Archaic Spartan Warfare"* in Eds: Luther, A., Meier, M. and Thommen, L. Das Frühe Sparta, 2006, pp. 125–164. Stuttgart: Franz Steiner.

Van Wees, Hans. *"Tyrtaeus' Eunomia: Nothing to Do With the Great Rhetra"*, Eds: Hodkinson, S. and Powell, A. Sparta: New Perspectives; Swansea: Classical Press of Wales, 1999, pp. 1–41.

Varro. *"On the Latin Language"*, trans. Kent, R. Loeb Classical Library. Cambridge, MA: Harvard University Press, 2014.

Vasaly, Ann. *"Characterization and Complexity: Caesar, Sallust and Livy"* in The Roman Historians. Ed: Feldherr, A. Cambridge: Cambridge University Press, 2009, pp. 245–260.

Vegetius. *"The Military Institutions of the Romans (De Re Militari)"* trans. Clarke, J. Ed: Phillips, T. Mansfield Centre, CT Connecticut: Martino Fine Books, 2011.

Wadge, R. *"King Arthur: A British or Sarmatian Tradition?"* Folklore, Vol. 98, No. 2 (1987), pp. 204–215.

Warren, Larissa Bonfante. *"Roman Triumphs and Etruscan Kings: The Changing Face of the Triumph"* The Journal of Roman Studies, Vol. 60 (1970), pp. 49–66.

Webster, Graham. *"Roman Imperial Army of the First and Second Centuries A.D."* London: Adam & Charles Black, 1969.

Weiß, Peter. *"Slingers' Lead Bullets"* in Brill's New Pauly, Antiquity volumes edited by: Hubert Cancik and Helmuth Schneider, English Edition by: Christine F. Salazar, Classical Tradition volumes edited by: Manfred Landfester, English Edition by: Francis G. Gentry. Leiden; Boston: Brill, 2006.

Whately, Conor. *"The War Cry: Ritulized Behavior and Roman Identity in Ancient Warfare, 200 BCE-400 CE"* in Imperial Identities in the Roman World. Eds: Vanacker, W. and Zuiderhoek, A. London: Routledge, 2016, pp. 61–77.

Whitby, Michael. *"Rome at War AD 293–696"* New York: Routledge, 2003.

Williams, Mary Frances *"Character of Aeetes in the Argonautica of Apollonius Rhodius"* Hermes, 124. Bd., H.4 (1996), pp. 463–479.

Williams, Thomas. *"A New Epithet of Mars"* Hermes, 93. Bd., H. 2 (1965), p. 252.

Wiseman, T. P. *"Calpurnius Siculus and the Claudian Civil War"* The Journal of Roman Studies, Vol. 72 (1982), pp. 57–67.

Wiseman, T. P. *"Practice and Theory in Roman Historiography"* History, Vol. 66, No. 218 (1981), pp. 375–393.

Worthington, Ian. *"Philip II of Macedonia"* New Haven, CT: Yale University Press, 2008.

Xenophon. *"Anabasis"*, trans. Brownson, C. Loeb Classical Library. Cambridge, MA: Harvard University Press, 2014.

Xenophon. *"Constitution of the Lacedaimonians"*, trans. E. C. Marchant, G. W. Bowersock, Harvard University Press, Cambridge, MA and London: William Heinemann, Ltd., 1925.

Xenophon. *"Cyropaedia"*, trans. Miller, W. Loeb Classical Library. Cambridge, MA: Harvard University Press, 2014.

Xenophon. *"Hellenica"*, trans. Brownson, C. Loeb Classical Library. Cambridge, MA: Harvard University Press, 2014.

Zhmodikov, Alexander. *"Roman Republican Heavy Infantrymen in Battle (IV-II Centuries B.C.)"* Historia: Zeitschrift für Alte Geschichte, Bd. 49, H. 1, 1st Qtr., 2000, pp. 67–78.

Zimmermann, Bernhard. *"Aeschylus"* in Brill's New Pauly, Antiquity volumes edited by: Hubert Cancik and, Helmuth Schneider, English Edition by: Christine F. Salazar, Classical Tradition volumes edited by: Manfred Landfester, English Edition by: Francis G. Gentry. Leiden; Boston: Brill, 2002.

Ziolkowski, John. *"The Invention of the Tuba (Trumpet)"* The Classical World, Vol. 92, No. 4 (Mar.–Apr., 1999), pp. 367–373.

Index

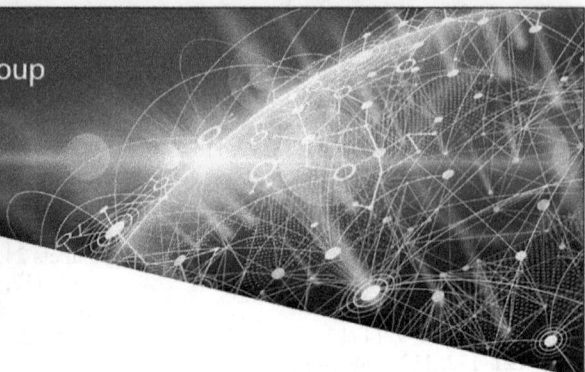

For Product Safety Concerns and Information please contact our EU
representative GPSR@taylorandfrancis.com
Taylor & Francis Verlag GmbH, Kaufingerstraße 24, 80331 München, Germany

www.ingramcontent.com/pod-product-compliance
Lightning Source LLC
Chambersburg PA
CBHW060313220326
41598CB00027B/4312

* 9 7 8 1 0 3 2 2 4 8 6 0 8 *